Reliability Centered Maintenance (RCM)

Reliability Centered Maintenance (RCM)

Implementation Made Simple

Neil Bloom

McGraw-Hill, Inc.
New York Chicago San Francisco Lisbon London
Madrid Mexico City Milan New Delhi San Juan
Seoul Singapore Sydney Toronto

The McGraw·Hill Companies

Library of Congress Cataloging-in-Publication Data

Bloom, Neil.
　Reliability centered maintenance : implementation made simple / Neil Bloom.
　　p. cm.
　ISBN 0-07-146069-1
　1. Plant maintenance. 2. Reliability (Engineering) 3. Maintainability (Engineering)
　I. Title.
　TS192.B56 2005
　658.2'02—dc22　　　　　　　　　　　　　　　　　　　　　　　　　2005054460

Copyright © 2006 by the McGraw-Hill Companies. All rights reserved. Printed in the United States of America. Except as permitted under the United States Copyright Act of 1976, no part of this publication may be reproduced or distributed in any form or by any means, or stored in a data base or retrieval system, without the prior written permission of the publisher.

1 2 3 4 5 6 7 8 9 0 DOC/DOC 0 1 0 9 8 7 6 5

ISBN 0-07-146069-1

The sponsoring editor for this book was Kenneth McCombs and the production supervisor was Pamela Pelton. The art director for the cover was Anthony Landi. It was set in Century Schoolbook by North Market Street Graphics.

Printed and bound by RR Donnelley.

McGraw-Hill books are available at special quantity discounts to use as premiums and sales promotions, or for use in corporate training programs. For more information, please write to the Director of Special Sales, McGraw-Hill Professional, Two Penn Plaza, New York, NY 10121-2298. Or contact your local bookstore.

This book is printed on recycled, acid-free paper containing a minumum of 50% recycled de-inked fiber.

Information contained in this work has been obtained by The McGraw-Hill Companies, Inc. ("McGraw-Hill") from sources believed to be reliable. However, neither McGraw-Hill nor its authors guarantee the accuracy or completeness of any information published herein and neither McGraw-Hill nor its authors shall be responsible for any errors, omissions, or damages arising out of use of this information. This work is published with the understanding that McGraw-Hill and its authors are supplying information but are not attempting to render engineering or other professional services. If such services are required, the assistance of an appropriate professional should be sought.

To my wife, friend, partner, and soul mate, Bernadette, who is indeed a saint

Contents

Preface xiii
Acknowledgments xix

Chapter 1 Introduction to RCM 1
1.1 Uncovering the Fuzziness and Mystique of RCM 4
1.2 The Background of RCM 9
1.3 A No-Nonsense Approach to RCM 11
1.4 RCM as a Major Factor in the Bottom Line 12

Chapter 2 Why RCM Has Historically Been So Difficult to Implement 15
2.1 Consultants 15
2.2 A White Elephant 16
2.3 Reasons for Failure 18
    2.3.1 Loss of In-House Control 18
    2.3.2 An Incorrect Mix of Personnel Performing the Analysis 19
    2.3.3 Unnecessary and Costly Administrative Burdens 20
    2.3.4 Fundamental RCM Concepts Not Understood 21
    2.3.5 Confusion Determining System Functions 21
    2.3.6 Confusion Concerning System Boundaries and Interfaces 21
    2.3.7 Divergent Expectations 23
    2.3.8 Confusion Regarding Convention 24
    2.3.9 Misunderstanding "Hidden" Failures and Redundancy 24
    2.3.10 Misunderstanding Run-to-Failure 25
    2.3.11 Inappropriate Component Classifications 25
    2.3.12 Instruments Were Not Included as Part of the RCM Analysis 26

Chapter 3 Fundamental RCM Concepts Explained, Some for the Very First Time: The Next Plateau 27
3.1 The Three Phases of an RCM-Based Preventive Maintenance Program 30
3.2 The Three Cornerstones of RCM 32
3.3 Hidden Failures, Redundancy, and Critical Components 34
3.4 Testing Hidden Systems 45

3.5	The Missing Link: Potentially Critical Components	46
3.6	Commitment Components	50
3.7	Economic Components	51
3.8	The "Canon Law" of Run-to-Failure Components	52
3.9	The Integration of Preventive and Corrective Maintenance and the Distinction Between Potentially Critical and Run-to-Failure Components	57
	3.9.1 An RTF CM versus a Critical CM: Which Takes Priority for Getting Worked First?	59
3.10	The Anatomy of a Disaster	61
3.11	A Deeper Look at Critical Components, Potentially Critical Components, and Hidden Failures—How They All Fit Together	65
3.12	Finding the Anomalies	68
3.13	Failures Found During Operator Rounds	70
3.14	Redundant, Standby, and Backup Functions	70
3.15	Typical Examples of Component Classifications	73
3.16	Component Classification Hierarchy	73
3.17	The Defensive Strategies of a PM Program	75
3.18	Eliminating the Requirement for Identifying Boundaries and Interfaces	75
3.19	Functions and Functional Failures Are Identified at the Component Level, Not the System and Subsystem Level	77
3.20	The Quest for the Consequence of Failure	79
3.21	The COFA versus the FMEA	81
3.22	How Do You Know When Your Plant Is Reliable?	83
3.23	Chapter Summary	85

Chapter 4 RCM Implementation: Preparation and Tools — 89

4.1	Preparation	90
4.2	The Sequential Elements Needed for the Analysis	91
	4.2.1 A Simple but Comprehensive Alphanumeric Equipment I.D. Database	91
	4.2.2 Informational Resources	93
	4.2.3 Establishing Convention	94
	4.2.4 Specialized Workstations and Software	94
	4.2.5 The COFA Excel Spreadsheet versus the FMEA	95
	4.2.6 The PM Task Worksheet	100
	4.2.7 The Economic Evaluation Worksheet	102
4.3	Chapter Summary	105

Chapter 5 RCM Made Simple: Implementation Process — 107

5.1	Define Your Asset Reliability Strategy	109
5.2	Understanding the RCM COFA Logic Tree, the Potentially Critical Guideline, and the Economically Significant Guideline	112
5.3	Completing the COFA Worksheet in Conjunction with the COFA Logic Tree, the Potentially Critical Guideline, and the Economically Significant Guideline	120
	5.3.1 Describe the Component Functions	121

			Contents	ix

	5.3.2	Describe the Functional Failures	123
	5.3.3	Describe the Dominant Component Failure Modes for Each Functional Failure	124
	5.3.4	Is the Occurrence of the Failure Mode Evident?	124
	5.3.5	Describe the System Effect for Each Failure Mode	126
	5.3.6	Describe the Consequence of Failure Based on the Asset Reliability Criteria You Selected	129
	5.3.7	Define the Component Classification	129
5.4		RCM Serves as a Translation of the Design Objectives	131
5.5		Companion Equipment	133
5.6		The SAE Standard: Document JA1011	134
5.7		A Real-Life Analysis: Averting a Potentially Devastating Plant Consequence	135
5.8		Why Streamlined RCM Methods Are Not Recommended	141
	5.8.1	Total Productive Maintenance (TPM)	142
	5.8.2	Reliability-Based Maintenance (RBM)	142
	5.8.3	Probabilistic Safety Analysis (PSA) Based Maintenance	142
	5.8.4	80/20 Rule	142
5.9		Chapter Summary	143
5.10		RCM Made "Difficult"	147
	5.10.1	Determine System Boundaries	148
	5.10.2	Determine Subsystem Boundaries	148
	5.10.3	Determine Interfaces	149
	5.10.4	Determine Functions	149
	5.10.5	Determine the Functional Failures	150
	5.10.6	Determine Which Equipment Is Responsible for the Functional Failures	150

Chapter 6 The PM Task Selection Process — 153

6.1	Understanding Preventive Maintenance Task Terminology	154
6.2	Condition-Directed, Time-Directed, and Failure-Finding Tasks	154
6.3	The PM Task Worksheet	157
6.4	The PM Task Selection Logic Tree	158
6.5	Why a Condition-Directed Task Is Preferred	161
6.6	Determining the PM Task Frequency and Interval	162
	6.6.1 The Optimum Time to Establish a Reliability Program	165
6.7	Is a Design Change Recommended?	166
6.8	Completing a Typical PM Task Worksheet	167
6.9	Institute Technical Restraints	168
6.10	A Sampling Strategy	169
6.11	Common Mode Failures	171
6.12	Different Predictive Maintenance (PdM) Techniques	172
	6.12.1 Vibration Monitoring and Analysis	172
	6.12.2 Acoustic Monitoring	173
	6.12.3 Thermography or Infrared Monitoring	173
	6.12.4 Oil Sampling and Analysis	173
	6.12.5 X-ray or Radiography Inspection	173
	6.12.6 Magnetic Particle Inspection	174

x   Contents

6.12.7	Eddy Current Testing	174
6.12.8	Ultrasonic Testing	174
6.12.9	Liquid Penetrant	174
6.12.10	Motor Current Signature Analysis (MCSA)	174
6.12.11	Boroscope Inspections	174
6.12.12	Diagnostics for Motor-Operated Valves	175
6.12.13	Diagnostics for Air-Operated Valves	175
6.13 Chapter Summary		175

Chapter 7 RCM for Instruments 181

7.1 Instrument Categories	182
7.2 Instrument Design Tolerance Criteria	183
7.3 The Instrument Logic Tree	185
7.3.1 Block 1: Is the Instrument a Functional Instrument?	185
7.3.2 Block 2: Instrument Is Analyzed in the COFA Worksheet and the PM Task Selection Worksheet.	185
7.3.3 Block 3: Can the Instrument Reading Result in an Operator Having to Initiate Some Kind of Action?	185
7.3.4 Block 4: A PM Is Required. Calibration Criteria and Periodicity Guidance Are as Follows.	186
7.3.5 Block 5: Were the Last Three Successive Calibrations Within Vendor Tolerance Criteria?	186
7.3.6 Block 6: Periodicity Extension Is Allowed.	187
7.3.7 Block 7: Reduce Periodicity or Implement a Design Change.	187
7.3.8 Block 8: Is the Instrument Redundant?	187
7.3.9 Block 9: Is an Indication Comparison Applicable?	187
7.3.10 Block 10: Is the Consequence of Excessive Drift (to the Point of Instrument Failure) Acceptable?	188
7.3.11 Block 11: A Calibration PM Is Optional.	189
7.3.12 Block 12: A PM Is Required. Calibration Criteria and Periodicity Guidance Are as Follows.	189
7.3.13 Block 13: Were the Last Two Successive Calibrations Within a +/−2.5 Percent Accuracy Tolerance?	189
7.3.14 Block 14: Periodicity Extension Is Allowed.	189
7.3.15 Block 15: Were the Last Two Successive Calibrations Within a +/−5.0 Percent Accuracy Tolerance?	189
7.3.16 Block 16: Periodicity Extension Is Not Allowed.	190
7.3.17 Block 17: Reduce Periodicity or Implement a Design Change.	190
7.4 Chapter Summary	190

Chapter 8 The RCM Living Program 193

8.1 A Model for an RCM Living Program	194
8.1.1 The Craft Feedback Evaluation Element	196
8.1.2 The Corrective Maintenance (CM) Evaluation Element	203
8.1.3 The "Other Inputs" Element	205
8.1.3.1 Root-Cause Evaluations	206
8.1.3.2 Vendor Bulletins	206
8.1.3.3 Regulatory Bulletins	207
8.1.3.4 Industry Failure Data	207
8.1.3.5 Engineering Evaluations	208

			Contents	xi
		8.1.3.6	Plant Design Changes	208
		8.1.3.7	New Commitments	208
	8.1.4	Monitoring and Trending		209
	8.1.5	The RCM Analysis Element		209
	8.1.6	Equipment Database		210
	8.1.7	The PM Audit		210
8.2	Chapter Summary			212

Chapter 9 An RCM Monitoring and Trending Strategy — 217

9.1	What Is Reliability and How Do You Measure It?	218
9.2	Monitoring Reliability Is Like Monitoring the Human Body	220
9.3	Caution: Avoid Analysis Paralysis Performance Monitoring	220
9.4	The Aggregate Metrics	222
	9.4.1 Unplanned Plant or Facility Trips	223
	9.4.2 Capacity Factor	224
	9.4.3 Unplanned Operator Actions	224
	9.4.4 Unplanned Power Reductions	225
	9.4.5 Production Delays	225
	9.4.6 Enforcement Actions	226
	9.4.7 Litigation Occurrences	226
	9.4.8 Citations and Violations	227
	9.4.9 Root-Cause Evaluations	227
	9.4.10 Injuries	228
	9.4.11 Rate of Written CMs	228
	9.4.12 Overdue CM Backlog	229
	9.4.13 Overdue PM Backlog	229
9.5	Weighting Factors	230
9.6	Performance Calculations	231
9.7	Performance Graph	235
9.8	Performance Graph by System	237
9.9	A Final Caution	239
9.10	Benchmarking	239
9.11	More About Expected Performance Rates	241
9.12	Avoid Reliability Complacency	241
9.13	How to Maintain Your Reliability Performance	242
9.14	Chapter Summary	246

Chapter 10 RCM Implementation Made Simple—Epilogue — 249

10.1	RCM as a Plant Culture	249
10.2	A Step-by-Step Review of the Process	251
	10.2.1 Select an RCM Point of Contact	251
	10.2.2 Review the Reasons for RCM Program Failures	255
	10.2.3 Understand the Concepts	255
	10.2.4 Define Your Asset Reliability Criteria	255
	10.2.5 Establish Your Alphanumeric Equipment Database	256
	10.2.6 Analyze Each Component Function in the COFA Logic Tree	256
	10.2.7 Analyze Each Component Function in the Potentially Critical Guideline	257

	10.2.8	Analyze Each Component Function in the Economically Significant Guideline	257
	10.2.9	Enter All Data in the COFA Worksheet	258
	10.2.10	Classify Each Component	258
	10.2.11	Analyze All Classified Components Except Run-to-Failure Components in the PM Task Selection Logic Tree	258
	10.2.12	Document All Tasks and Periodicities on the PM Task Worksheet	258
	10.2.13	Analyze Instruments in the Instrument Logic Tree	259
	10.2.14	Develop Your RCM Living Program	259
	10.2.15	Establish Monitoring and Trending Program Metrics	260
	10.2.16	Establish Your Expected Performance Rate	260
	10.2.17	Establish Your Actual Performance Rate	261
	10.2.18	Establish Your Trend Graphs	261
	10.2.19	Maintain Continued Vigilance Over Your Program	261
10.3	Taking Command of Your Own Ship		262

Glossary   265
Bibliography   285
Index   287

Preface

Reliability centered maintenance, or RCM, as it is called, *was* difficult. RCM *was* an albatross. It *was* cumbersome, expensive, and almost impossible to implement. Note my use of the past tense here! Implementing an RCM program has for the most part been shrouded in confusion, and its image has taken on an aura of perceived complexity. I plan to change that.

I am writing this book because I find that most other books on this subject are very difficult to understand and even more difficult to use as a tool for implementation. RCM is a very powerful reliability tool, but as long as it remains non-user-friendly, its full potential is limited. It is my belief that classical RCM has been made much more complicated than it needs to be. I explain classical RCM (not streamlined RCM) in simple terms and introduce some new concepts that have never before been identified. You will learn how to readily implement an affordable premier reliability program for your plant or facility, on your own, without the need for any outside expertise and without the need for special training of any kind. I truly believe that this book has the potential to set a new standard for preventive maintenance and reliability via the classical RCM process.

You are probably asking yourself... "Who is this author, and how can he explain how to implement classical RCM in a simple, straightforward manner, easily understandable to nontechnical as well as technical personnel the world over?"

I have been responsible for developing and managing what is perhaps, even today, one of the most comprehensive classical RCM programs ever implemented. The program analyzed every system, covering more than 125,000 individual components, at one of the country's largest nuclear generating facilities. Some of the ideas and concepts that I developed in 1991 are now specifically documented in the latest SAE Standard for RCM published in September 1999.

My Background

I received my bachelor of science degree in mechanical engineering from the University of Miami, where I also minored in economics. I have been a guest speaker on RCM at some of the most prestigious national and international conferences. These include those of the Electric Power Research Institute (EPRI), American Society of Mechanical Engineers (ASME), Edison Electric Institute (EEI), Argonne National Laboratory (ANL), which is operated by the University of Chicago for the Department of Energy (DOE), the American Nuclear Society (ANS), and the International Atomic Energy Agency (IAEA) in Vienna, Austria. The Canadian Atomic Energy Control Board (AECB) requested that I personally meet with some of its members to discuss the RCM program that I had developed.

My engineering and maintenance career of almost 40 years has been devoted to the commercial aviation and commercial nuclear power industries. Both of these require the highest standards of safety and reliability, as evidenced by their highly stringent regulation by the federal government via the Federal Aviation Administration (FAA) and the Nuclear Regulatory Commission (NRC). I have been fortunate to have worked closely with both of these entities.

In 1967, I began working as a systems engineer at one of the nation's largest airlines, which had more than 30,000 employees and a fleet of several hundred aircraft. I progressed to superintendent of intermediate aircraft maintenance and then became the administrative assistant to the vice president of maintenance.

My experience included Maintenance Steering Group (MSG-2 and MSG-3) reliability studies in which MSG logic was the forerunner to RCM. Working closely with aircraft manufacturers and their suppliers to enhance safety and reliability objectives, I was instrumental in establishing aircraft maintenance strategies, initiating aircraft design changes, and interfacing with the

FAA as a liaison. I developed aircraft preventive maintenance programs—from the Douglas DC-8/DC-9 and Boeing B707/B727 to the Airbus A300, Lockheed L1011, and the Boeing B757—and was also a member of the Maintenance Review Board (MRB) for the Lockheed L1011. Commercial aviation is where RCM was first introduced. It made its way to the nuclear power industry in the mid-1980s.

Beginning in 1983, I worked for one of the nation's largest electric utilities at its nuclear generating facility. I was involved with NRC regulatory issues; maintenance engineering activities; maintenance procedures, policies, and practices; and, from 1991 until my retirement in 2004, I was the program manager responsible for RCM and preventive maintenance programs.

Commercial aviation and nuclear power, paramount in the hierarchy of safety and reliability relative to most other industries, have afforded me the special practical experience and expertise to know what can and cannot be done with classical RCM. I know what works and what doesn't work, what the pitfalls are, and how to circumvent the roadblocks. I know what changes can be made to maintain the same, or even more, robustness of the process while minimizing the administrative burdens. I know what information is absolutely necessary to implement a successful program, and how to do this with ease. I also know what parts of the process are not necessary and do not need to be included.

Everything I explain in this book is in total accord with the original airline MSG and RCM methodology and the latest SAE Standard governing RCM (designated as JA1011), which I discuss in great detail in Chapters 3 and 5. In fact, some of the very specific ideas in the RCM program I developed in 1991 are now included in the new JA1011 SAE RCM Standard titled *Evaluation Criteria for RCM Processes*.

My wife and I live in Monarch Beach, California. You may contact me via e-mail at neilbloom@rcmauthor.com.

Some Additional Insight from the Author

I would like to mention the work of a colleague of mine, John Moubray, who recently passed away. I first met John when he came to visit me in California in early 1991 after becoming aware of my work on RCM in the nuclear industry. He was an outstanding advocate of RCM, and his efforts helped to bring it the visibility it justly deserves. Like John, I, too, am an advocate of classical RCM versus the shortcut versions, but I believe classical RCM can be achieved with a much more simplified approach.

Finally, it was from comments I received after having given a presentation at the Southern California Plant Engineering and Facilities Maintenance Conference that reinvigorated my reasons for writing this book. Many people from relatively small and midsize industries and facilities came up to me afterward and told me they wanted to implement a classical RCM program but that their companies did not have hundreds of thousands of dollars to spend on the program—and certainly not millions of dollars, like the nuclear industry does. They wanted the same rigorous analysis, but they did not believe they had the knowledge or the financial resources to implement it. They wanted a book that could guide them through the RCM process without having to spend large sums of money for consulting expertise and without having to rely on hundreds of engineering and other technical personnel who were not available to them.

Since my retirement in 2004, I have devoted my full efforts to writing this book, which has been almost 14 years in the making. My goal is to enable and empower you to implement a premier classical RCM program at your facility without having to spend an inordinate amount of time and money and without the need for expensive outside consulting services, specialized training, or other support. I have embraced straightforward, easy-to-understand logic, have used an objective rather than a subjective decision-making process, and have given great importance to maintaining the conceptual clarity of the process to highlight its simplicity. These are all designed specifically to enhance the understanding, implementability, and cost-effectiveness of RCM. When you have finished reading this book you will be able to establish an affordable and robust premier reliability program that will make your facility safer, more reliable, and more cost efficient.

This could be the RCM breakthrough that you have been looking for, and I hope that you will find each of the following chapters to be a revelation. It is my belief that industry, universally, has the potential for attaining even greater levels of safety and reliability if the RCM process becomes more user-friendly, as it was intended to be by its pioneers, Stanley Nowlan and Howard Heap.

Acknowledgments

I would like to acknowledge all of my peers, subordinates, and superiors in the airline industry who gave me the opportunity to learn the valuable importance of safety and reliability. I would especially like to thank Eldon Johnson, of Pan American World Airways, and Sam Miller, Jack Steffen, and Fred Lind, former vice president of maintenance at Eastern Airlines.

Primarily, I would like to acknowledge the work of Stanley Nowlan and Howard Heap of United Airlines, the true pioneers of RCM, whose treatise on airline RCM principles is still the primary basis for this preventive maintenance program methodology.

I am indebted to many people in the nuclear industry for having the wisdom and the vision to allow me to foster the development of classical RCM from its airline origins. I would especially like to thank Hans Merten, Steve McMahan, and vice presidents Joe Wambold and Russ Kreiger of Southern California Edison.

I would also like to thank the Electric Power Research Institute (EPRI) and the Institute of Nuclear Power Operators (INPO), from whom I gained great insight into the culture of safety and reliability. I would specifically like to mention John Gaertner of EPRI for his invaluable friendship through the years.

A special acknowledgment goes to Bob Baldwin of *Maintenance Technology* magazine and to the Society of Maintenance and Reliability Professionals (SMRP) for their unrelenting efforts to bring reliability out of the dark ages and into a leading-edge role for furthering an industrial safety and reliability culture and as a means for enhancing corporate profitability.

Finally, I want to acknowledge many unknown individuals. These were the attendees at various conferences in which I was a guest speaker on RCM. Many of these individuals represented rather small facilities compared to nuclear power plants or fleets of commercial aircraft. They inquired how they could learn to develop and implement a premier RCM program without having the deep pockets to do it with. That was, perhaps, the final inspiration I needed to begin this journey.

Finally, I want to express my total appreciation to Ken McCombs, my editor at McGraw-Hill, and most of all to my wife Bernadette, who so patiently provided the inspiration and the environment for this book to come to fruition.

Chapter 1

Introduction to RCM

Reliability centered maintenance (RCM) is not new. Airline Maintenance Steering Group (MSG) logic, the predecessor to RCM, has existed since the early 1960s. Stanley Nowlan and Howard Heap of United Airlines introduced formal RCM to the commercial aviation industry in 1978. Airline preventive maintenance and reliability is primarily based on their work, and they are considered to be the "grandfathers" of RCM. Their vision is as relevant today as it was when they published the first (and most authoritative) rendition of *Reliability Centered Maintenance* in 1978.

RCM is nothing more than a logical way of identifying what equipment in your facility is required to be maintained on a preventive maintenance basis rather than a let-it-fail-then-fix-it basis, commonly referred to as *run-to-failure* (RTF). Many of you have heard the phrases "don't fix it until it breaks" or "don't break it by trying to fix it." There is a grain of truth to these axioms, but they depict a very shallow approach if you are striving to achieve reliability and safety levels for your facility that are the best they can be.

Many plants and facilities have tried the hit-and-miss approach, or the old "how-we-used-to-do-it" approach, or the run-on-luck approach to maintenance. These methods will get you only so far until your luck runs out, and the potential for disaster looms right around the corner. In the absence of a structured RCM approach, reliability will rest solely on the basis of seat-of-the-pants experience, with a strategy consisting of a best-guess decision process. That approach falls far short of modern-day expectations.

Disasters can be caused by acts of nature, human error, or equipment failures. Disasters caused by acts of nature such as hurricanes, earthquakes, tsunamis, tornados, or landslides do not lend themselves to being tamed by human intervention; for the most part, they are unavoidable. There may be warning systems, such as tsunami warning buoys in the Pacific Rim, or construction standards that help to prevent structures from buckling during an earthquake, but the event itself is unavoidable. On the other hand, human error (pilot error, judgment error, operator error, etc.) offers a range of latitude in circumventing the potential for disaster. Some of the tools that might be used, for example, include better training, more specific procedural guidance relative to performing a given task, a safer work environment, and more rigid standards and codes—all actions that can be taken to avoid human error to some degree.

What about disasters that happen in factories, plants, or other facilities that usually have their origin in the failure of equipment? These types of failures probably offer the greatest latitude of all for circumventing their potential to cause a disaster. Nothing is ever 100 percent reliable, whether it is an aircraft, a space shuttle, or a nuclear power plant. However, disasters caused by equipment failure have the capability to be harnessed to the degree that allows for the closest proximity to that 100 percent reliability threshold. That cannot be said for natural disasters or for human-induced ones.

We have a lot of control over the way we maintain our facilities and equipment to prevent failures. A reliability centered maintenance approach to preventive maintenance is probably the best path you can take to get as close as possible to that 100 percent reliability threshold. An RCM analysis also considers the fact that maintenance budgets are not unlimited, and thus some rational basis exists for deciding what to do and where to expend the most effort.

Today, almost everyone in a manufacturing, power generation, production, and other technological environments is familiar with the terminology of RCM. However, one's perceived degree of familiarity may be quite deceiving. RCM is very simple in concept but also very sophisticatedly subtle in its application. As with many processes, having only a very limited understanding of the RCM process may, in fact, prove to be more problematic than beneficial. The false comfort level of naively believing that a superficial implementation of the process will be a panacea for plant and equipment problems, and then depending on that process to produce significant reliability results, is unrealistic.

The understanding of RCM that many of us have comes from reading books and articles on the subject or from consultants' sales presentations. Oftentimes this information is limited and includes terminology such as *boundaries, functions, interfaces, functional failures,* and so on. More specifically, the terminology should include: *establishing system boundaries, subsystem boundaries, in-system in-interfaces, in-system out-interfaces, out-system in-interfaces, out-system out-interfaces, system functions, subsystem functions, failures of those subsystem functions, consequences of those functional failures,* and so on. The very mention of these phrases have probably caused an immediate quizzical look on your face, and rightly so. These are some of the very reasons why RCM has been so difficult to implement. Chapters 2 and 3 explain why much of this entire cumbersome process is not even needed.

A more meaningful understanding of reliability can quickly become apparent thorough reflection on some of the concepts I introduce in this book. I developed these concepts because of the difficulty I witnessed time and again by midlevel and senior-level management types trying to understanding the RCM process. It is not rocket science, but it is sophisticated and subtle.

Recognizing hidden failure modes, understanding when a single-failure analysis is not acceptable, and knowing when run to failure is acceptable are the real cornerstones of RCM. Additionally, the understated, but powerfully important, distinction between true redundancy and redundant components fulfilling a standby or backup function is a key to reliability success.

Although Stanley Nowlan and Howard Heap gave great importance to the principle of "hidden" failures, unfortunately, hidden failures are not widely understood and are often overlooked. Almost everyone has heard the terminology *hidden failure*. What I have found, however, is that very few people understand the difference between a redundant system, a backup system, and a standby system and how hidden failures affect them. It is not well understood that different subtleties, which may appear to be identical operating conditions, can cause very different outcomes that result in a component being classified as either *immediately critical* or *run-to-failure.*

Many utilities and other industries have implemented various forms of an RCM program only to find that they continued to have fundamental reliability issues that were not addressed by their analysis. The primary reason is the lack of a grassroots philosophical understanding of the principles governing the analysis.

1.1 Uncovering the Fuzziness and Mystique of RCM

In my presentations at various maintenance and engineering conferences around the country, I have found that there is a certain "fuzziness," or mystique, associated with implementing an RCM program. Much of this fuzziness seems to arise when attempting to analyze redundant equipment, identify hidden failures, invoke a run-to-failure strategy, determine when a single-failure analysis is acceptable, and decide when (if not why) a multiple-failure analysis is required. I clearly delineate the differences between these terms and at what point each one is applicable. In Chapter 3, I explain my concept of *potentially critical components,* my *"canon law" for run-to-failure,* and how to differentiate when a single-failure analysis can be used versus a multiple-failure analysis. You will also be introduced to the *consequence-of-failure analysis* (COFA). You will find these concepts to be extremely powerful tools; however, they nonetheless are quite simple and straightforward.

These concepts have been made easy to understand, since they form the very fabric for successfully implementing an RCM program. These are concepts I have created that directly descend from, yet go beyond, the work of Stanley Nowlan and Howard Heap. These concepts have not been clearly espoused prior to my writing this book, which is why it is titled *Reliability Centered: Maintenance Implementation Made Simple.*

In the past several years numerous versions of the RCM process have evolved. They have been called *streamlined, abbreviated, shortcut, truncated,* and so on. It is this author's belief that there is only one real RCM process, and that is the *classical* RCM process. Other, truncated, versions evolved only because of the difficulty and expense in attempting to implement the classical version. It is also my belief that these other versions will not afford you a comprehensive reliability program, because many important functions and potential failure consequences will be missed. In fact, serious mishaps may even occur as a result. I discuss, in detail, the shortcomings of streamlined versions of the process in Chapter 5. Throughout this book, whenever I refer to RCM, it is the *classical* version I am referring to unless otherwise noted.

Being familiar with the shortcut approach and the classical approach, I compare it to the following analogy. . . . Suppose a person smoked two packs of cigarettes per day. Cutting down to one pack per day would provide a better outcome. However, if that person cut down to zero packs per day, that would offer an even more optimum outcome. Now, if cutting down to zero packs per day required virtually the same effort as cutting down to one pack per day, or possibly even less effort, which would you think is best? While streamlined RCM is better than not having any basis for your preventive maintenance program, it is still tantamount to a pick-and-choose potpourri philosophy, and streamlined versions will not uncover those innocuous challenges that potentially jeopardize your plant or facility. An abbreviated RCM process cannot provide for that optimum outcome, like the zero-pack-per-day smoker has optimally improved his results. The approach to classical RCM that you will be learning provides a greater opportunity to achieve that optimum outcome and, surprising as it may seem, will not take any more time or expense; in all likelihood, it will take even less effort than the shortcuts.

Remember, it is not the obvious that creates the greatest potential for disaster . . . it is the nonobvious! Streamlined RCM will not robustly ascertain the nonobvious consequences of failure. Only classical RCM offers the opportunity to find those relatively unknown, and what may appear to be noncritical, components, that can, in fact, have some of the most significant consequences as a result of their failure. You will see how classical RCM can be achieved in virtually the same amount of time, or less, than streamlined versions. This is discussed in detail in Chapters 3, 4, and 5.

Some RCM books and other RCM guides may lead you to believe that what appear to be "less-important" systems and components like service water, for example, are automatically run to failure without any further analysis required, such as might be the case with those espousing a streamlined RCM approach. Nothing can lead you further away from achieving reliability than to go off in that direction. As a case in point, for many years nuclear components were considered to be either safety-related or non-safety-related and were identified by their *quality class,* which ranged from Q-class 1 to Q-class 4 depending on whether they were part of the safety-related nuclear steam supply system (NSSS) or the non-safety-related balance-of-plant (BOP) system.

Having come from an airline background prior to my nuclear experience, I knew that every component needed to be treated equally, with the analysis laser-focused on the question *"what is the consequence of failure?"* regardless of any preconceived pedigree of the relative importance of a given system or component. The RCM program I developed in the early 1990s did not use the generic safety-related versus non-safety-related litmus test, but instead zeroed in on the consequence of failure regardless of any other pedigree placed on the component. This was a major departure from the existing nuclear thought process at that time. Then, in the late 1990s, to its credit, the Nuclear Regulatory Commission (NRC) established subsequent guidelines that required non-safety-related components to be analyzed for their consequence of failure. Even today, many reliability engineers and other professionals still do not comprehend the rather simple axiom that all components are assumed to be important until proven otherwise via a comprehensive analysis.

To further illustrate my point, the service water system at a plant may, at first glance, easily lead one to believe it supplied service water only to the lavatories and water fountains and is therefore unimportant and probably a run-to-failure system. The primary source of service water is usually the local water district. However, when looking at a certain service water schematic in detail, it was identified that there was a single check whose function was not only to ensure the continued supply of service water to the lavatories and water fountains, but *also* to provide the path for seal water flow to the bearings of all eight condenser circulating water pumps in the event of a city water-line break. There are four circulating water pumps per unit, meaning that the failure of one innocuous component in the rather nondescript service water system, under the right (or, more appropriately, *wrong*) conditions, could possibly result in a dual-unit shutdown. Employing streamlined techniques and making the wrong assumptions would have missed this critical function entirely. The specific component within the service water system was therefore *critical* and not run to failure.

It is important that no system be automatically discounted from the analysis. While I am not saying that, in some instances, in-house experience may appear to justify discounting a system, it should be done only after an analysis has been performed to make sure something has not been overlooked. Every component (switch, pump, valve, motor, etc.) has to be looked at. Each component was included in the design of your facility for a reason. Otherwise, why is it there? Remember, as I discuss in Chapter 3: It is not the obvious that everyone knows about, but rather it is the nonobvious failure that poses the most disastrous threat to your facility. If you don't analyze each component individually, you will miss the opportunity to identify those nonobvious failure modes.

Some RCM practitioners have even espoused an 80/20 rule, or some facsimile, that looks at only 20 percent of a plant and ignores the other 80 percent. In doing this, you most likely will not achieve the reliability goals you are looking for. Those innocuous components whose functions may be thought to be unimportant may in fact have unanalyzed hidden failure consequences that will remain in a "sleeper cell" mode just waiting to manifest themselves and wreak havoc on your facility. In Chapter 3 you see how these sleeper cell failure modes really work and how easy it is for them to bring a calamity to your doorstep, like the Trojan horse in Greek mythology.

I have even seen where the 80/20 rule was predicated on the number of corrective maintenance (CM) events. Because of the inaccuracies of arbitrarily counting the number of corrective maintenance orders without regard to the relative importance of those CMs, this would be a totally misleading criteria. A very critical component may have very few CMs, whereas a much less critical component may have many CMs. Again, not a very good way to identify an RCM population of important components.

A leading newspaper recently headlined a rather unpleasant, but poignant, example of how commencing an RCM analysis without understanding the concepts that you will be learning about in this book can indeed result in a disastrous outcome. An apparently flawed RCM attempt at a major theme park of a world-renowned corporation resulted in unwanted international attention and publicity because of mechanical equipment problems with one of its ride attractions that ultimately resulted in a fatality. The concepts of run-to-failure, redundancy, and potentially critical components were apparently not well understood.

RCM has three phases to it. Oftentimes these are combined, but that only creates another source of confusion. The first phase, which is the most important, is to identify the equipment that requires preventive maintenance. The next phase is to specify the different types of preventive maintenance activities and tasks, including predictive maintenance (PdM) techniques that need to be performed on the identified equipment. The third phase is ensuring that the preventive maintenance tasks that were specified are properly executed in a timely manner. These separate and discrete phases are discussed in Chapter 3.

Let's start with the definition of reliability centered maintenance (RCM):

> A set of tasks generated on the basis of a systematic evaluation that are used to develop or optimize a maintenance program. RCM incorporates decision logic to ascertain the safety and operational consequences of failures and identifies the mechanisms responsible for those failures.

This may seem like a lot of words to describe a simple, straightforward, logical process. How and why did this logical process come into being?

1.2 The Background of RCM

In the early years of commercial jet aviation, the aircraft manufacturers and the individual airlines believed that if an aircraft was overhauled at a given time interval and completely torn apart, virtually system by system, component by component, once it was released from the hangar it would perform totally reliably until the next major overhaul, notwithstanding the requisite intermediate maintenance activities required. Most of the equipment was completely overhauled whether it needed it or not. What the aircraft manufacturers and their customers (i.e., the airlines) found out was that expected levels of reliability were still elusive. Therefore, they believed that if they performed this overhaul more often, surely the reliability levels they were seeking would be achieved. Consequently, the overhaul periodicity was decreased. Note that I used the word *periodicity,* not *frequency.* I do this in order to be technically correct. The periodicity includes the frequency *plus* the interval. For example, an A2 periodicity includes the annual frequency "A" *plus* the interval of "2," meaning the task is performed every two years. Using the frequency alone can be misleading. For example, extending the frequency from weekly to monthly means you perform the task *less often,* thereby increasing the periodicity. Conversely, reducing the frequency from monthly to weekly means you perform the task *more often,* thereby decreasing the periodicity.

Once again, an entire aircraft and virtually all of its components were completely torn apart at a lesser periodicity (more often), and again it remained in the hangar for weeks earning no revenue. Once released from the hangar, expected levels of reliability were still not achieved and in fact were even less than expected. This anomaly created the environment that set up the priority for, and led to the work of, Nowlan and Heap. They began to understand that preserving critical equipment functions rather than randomly and arbitrarily tearing an entire aircraft apart was the key to reliability. They also found out that indiscriminately overhauling equipment actually had a reverse negative effect on reliability, because the probability of failure of the newly replaced equipment increased due to premature failures and infant mortality.

Another interesting phenomenon they found was that similar components did not wear out over time in any sort of identical manner. In fact, Nowlan and Heap showed that only approximately 11 percent of all components exhibited a wear-out rate that lent itself to replacement at a given periodicity. That meant that almost 90 percent of all other components failed randomly. Scheduled overhauls were therefore counterproductive for this population. This is discussed in detail in Chapter 6.

It was also recognized that a maintenance program cannot correct deficiencies in the inherent safety and reliability levels of the equipment. It can only prevent deterioration of those inherent levels. If the inherent reliability levels are found to be unsatisfactory, a design modification may be necessary to obtain any further improvement.

I would also like to note that the RCM process must remain a "living" one. It is never static. New failure modes may become evident, and additional information relative to equipment performance may present itself at any time. Oftentimes, scheduled periodicities of certain PM tasks may need to be adjusted. Periodicities may need to be increased or decreased. Newly identified tasks may need to be added, while others may need to be deleted based on new or different operating conditions or plant modifications. In Chapter 8, I show you how to establish a very effective but simple "living program." It will include a craft feedback loop that I have found to be an extremely important part of the living process because it helps to maintain the viability of the program. I also show you how to establish a "monitoring and trending program" that incorporates an aggregate of parameters and criteria to monitor the *effectiveness* of your RCM program. This program is thoroughly discussed in Chapter 9.

While RCM had its origin in commercial aviation, remember that it is a *universal* reliability process that is just as applicable for a shoe factory as it is for a nuclear power plant or a commercial jet aircraft. An effective RCM process will allow your preventive maintenance program to evolve from a level based primarily on vendor recommendations, random selection, or arbitrary assignment to one based on more prudent fundamentals such as a component functional analysis and the identification of any subsequent safety or operational consequences to your facility as a result of the component functional failure. This will provide greater confidence that your preventive maintenance program consists of only those tasks that are specifically required for the safe, reliable, and efficient operation of the plant and that any unnecessary work has been eliminated.

Scheduling unnecessary PM activities may actually result in a diminution of overall plant reliability by virtue of the burdens it places on both operations and maintenance personnel. These additional burdens, which include hanging and removing equipment tags, providing clearances, tracking PMs, monitoring work activities, and so on, all contribute to the unnecessary depletion of available resources.

1.3 A No-Nonsense Approach to RCM

This is a no-nonsense book in the sense that I do not dwell on the laws of physics and thermodynamics, the metallurgical properties of materials, probabilistic studies, Poisson distributions, the theoretical reasons for having a preventive maintenance program, or the in-depth history of society and its relationship with preventive maintenance. I also intentionally avoid any other esoteric information that, in my opinion, does not directly foster the simplified understating of RCM.

Rather, this book is intended for those people responsible for ensuring the reliability of their plant or facility who want to readily use this information to develop a premier reliability program based on the principles of RCM. It is intended to be a how-to book for those who want to implement an affordable program without the need for outside consultants and without the implicit requirement of having to obtain an engineering degree to understand and speak the language. The only peripheral reference to commerce that I mention is how reliability directly affects the corporate bottom line.

1.4 RCM as a Major Factor in The Bottom Line

Today's corporate world employs some of the most sophisticated strategies ever assembled for achieving business success, and one of those strategies is relatively new to industries outside of commercial aviation and nuclear power. I call this process an *asset reliability strategy*. I explain this in detail in Chapter 5.

Asset management is one of many corporate buzzwords used today. Some previous buzzwords were *synergy, synergistic optimization, cost containment, strategic assets, strategic planning,* and so on. These buzzwords and any such future buzzwords have one thing in common: They all depend on preserving corporate assets. More than ever, the bottom line of a corporation is dependent on the *reliability of its output.* By that I mean reliability levels that minimize any unplanned production delays, maintain generation capacity, ensure personnel and plant safety, and prevent any unwanted regulatory or environmental issues from bringing unwanted publicity and/or litigation. In essence, asset management, or whatever it may be called in the future, relies on an RCM approach as the core for identifying equipment functions that must be preserved to protect the corporate assets and ensure the uninterrupted and continuous corporate revenue income stream.

No longer is the maintenance organization relegated to second-team status behind sales, marketing, and finance. There is definitely a culture shift taking place within the universal industrial complex that is elevating the importance of the maintenance organization as part of the corporate flagship team. After all, where would these industrial corporations be without a first-class preventive maintenance program to ensure the reliable operation of the facility? When I talk about a plant or facility, I am talking about any type of plant or facility, whether it is a private or a public entity. It could be a power-generating facility, the electrical power transmission and distribution network, a shoe factory, a chip maker, a computer manufacturer, a copper mining facility, an oil refinery, an offshore oil platform, a daily newspaper, a paper mill, an automotive assembly line, a missile or armament production facility, the space shuttle, the aerospace industry, a military defense manufacturing plant, a hospital, a cruise ship, a chemical plant—in other words, any entity that manufactures a product or produces an output where it is unacceptable to incur unplanned interruptions of the operation or, worse yet, an unwanted disaster.

I am writing this book to be used universally: It is applicable for any type of industry, any size industry (large or small), and for any number of reliability-type employees. One extreme, of course is a nuclear power plant with several hundred engineering, maintenance, and operations personnel responsible for reliability. The other extreme is a small manufacturing facility with only a handful of personnel responsible for reliability. The RCM concepts and implementation process that I will lead you through are the same.

I also intend for this book to be a source of information for engineering and business management students who should at least have a working knowledge of real life plant and equipment reliability principles to go along with the theory they learn, since bottom-line corporate profits are directly affected by how reliably a facility is maintained.

I explain why RCM has had such a sordid history of being so difficult to implement. Heretofore, the knowledge required for implementing a program was kind of esoteric, and many believed that it required the expertise of consultants. I intend to remove that shroud of complexity so that anyone with just average technical intelligence can implement a premier RCM program and be able to understand all of the fundamentals of the process.

I have designed this book to be a completely comprehensive and self-contained guide to RCM, not only from the analysis point of view, but also by identifying the pitfalls to avoid, by introducing new concepts that make RCM simple, by discussing the uncomplicated tools you will need to commence the analysis, by explaining the step-by-step RCM implementation logic process with actual real-life examples and the step-by-step explanation of the preventive maintenance (PM) task strategies, by explaining how to establish an RCM living program, by explaining how to monitor and trend the performance of your RCM program, and by knowing when you have achieved the optimum balance point of your RCM efforts. All of this information is sequenced as follows:

- Chapter 2 explains the pitfalls of RCM and how to avoid them.
- Chapter 3 identifies the concepts of RCM and how to apply them.
- Chapter 4 identifies the tools needed to commence the analysis.
- Chapter 5 explains the step-by-step RCM analysis logic process, including each question of SAE Document JA1011.
- Chapter 6 explains the step-by-step PM task selection process.
- Chapter 7 explains how to implement RCM for instruments.
- Chapter 8 explains how to establish an RCM "living program."
- Chapter 9 explains how to establish a monitoring and trending program to monitor the effectiveness of your plant performance.
- Chapter 10 discusses RCM as a plant culture.

We begin by trying to understand why RCM has been so difficult to implement.

Chapter

2

Why RCM Has Historically Been So Difficult to Implement

It has been estimated that more than 60 percent of all RCM programs initiated have failed to be successfully implemented. Many of the other 40 percent that were completed were performed quite superficially, making their true value only marginal. Why has it been so difficult? Why has its success been so elusive? There is a reason. In fact, there are many reasons. You will learn what the pitfalls to success are, why they happen, where they happen, and how to avoid them. It is my goal to take the mystery out of the process so that it can be readily understandable and easily implemented. In my opinion, RCM has become overly complicated in its transfer from the airline industry. It is also my belief that the successful implementation of the process is inversely proportional to the complexity it has acquired.

2.1 Consultants

RCM has become, and still is to a large degree, a cottage industry for consultants. It is unfortunate, but true, that within the world of consultants, oftentimes they will either know less than you do, or if they do know more, their methods may employ an elixir of obfuscation to allow them sole possession of understanding the process, and hence a continued income stream.

Things are not usually that complicated, and RCM is a prime example. While I am not discounting the need for consultants, they should primarily be brought in as a temporary augmentation of your staff to assist in establishing a program under your direction. I cannot overemphasize the importance of maintaining in-house control, responsibility, and ownership of the process. To further this thinking, most RCM "consultants" I have met have never personally implemented a comprehensive classical RCM program. I have seen so many instances where someone had read something about RCM and became versed in the catchwords and catchphrases and instantly became a self-anointed consultant.

2.2 A White Elephant

When RCM came on the scene in industries other than commercial aviation, it did not take long for the white elephant stigma to be placed on it. This was particularly distressful because the reliability enhancements that should have been so easily attainable were instead held in abeyance because the process could not be implemented effectively. Many consultants, being unfamiliar with the airline model, did not fully understand how to implement an RCM analysis, and the cost of a full-blown effort at a nuclear plant, for example, reached into the millions of dollars, most often with very few tangible results to show for it. Some utilities went through this iteration several times. It was touted as a cost-saving preventive maintenance (PM) reduction effort in order to ingratiate the consultants with upper management. If they came in with the message that they could make a plant more reliable but it would result in added costs and increased personnel, what kind of reception do you think they would receive?

I want to state very clearly that RCM is *not* a PM reduction program. It is a reliability program. The results of an RCM analysis are what they are. There is no bias to either delete work or add work. If your facility is one that is laden with an inordinate number of PM tasks, many of which are believed to be unnecessary, RCM will indeed identify those unnecessary PMs and they will become candidates for deletion. On the other hand, if the PM program at your facility is a Spartan one, the process will probably add PMs to your program, but they will be PMs that weren't being done that should have been done. My experience has shown that although it is definitely beneficial to delete unnecessary work, the true benefits of an RCM program are not measured by the work that can be deleted. Instead, it is more accurately measured by some of the tasks added to the preventive maintenance program that were not being done prior to the analysis but should have been.

If your facility first established its PM program based on performing every task specified in each vendor manual, you will undoubtedly have too many PMs and you will be afforded the opportunity to delete a rather large number of the unnecessary ones. Many facilities base their PM program on the experience of some of their older employees. While this is commendable, it is not, however, in itself a valid basis on which to stake the reliability of your plant.

Nowlan and Heap wrote the original RCM treatise in aircraft terminology using examples found in commercial aviation. The aircraft language, as well as the process itself, resulted in significant confusion by those trying to transfer it to other industries, and they left out some of the most important aspects along the way. With extensive practical experience in airline and nuclear reliability and maintenance, I show you how to circumvent the mountains of misunderstanding and confusion that have regrettably found their way into current RCM adaptations.

RCM is almost always described as a process of identifying critical components whose failure would result in an unwanted consequence to one's facility. What if the component failure does not cause an immediate unwanted consequence? Is that automatically a noncritical component? Emphatically *no!* If there is built-in system redundancy, will that automatically allow a component to be run to failure? Again, emphatically *no!* Are run-to-failure components *un*important? Once again, emphatically *no!* In virtually all of the books on RCM there have been a myriad of "dots" surrounding these aspects. The dots were all there, but they were quite difficult and confusing to connect. I will show you how to connect those dots and build a bridge to a very straightforward process.

One of the keys to this is my concept of *potentially critical* components. I created this concept to solve the *missing link* of RCM. I write a lot about this vital concept and several other concepts in the next chapters because they are extremely important building blocks leading to the understanding of what RCM is all about. This will help guide you in forging a path at your facility toward a clear and concise understanding of RCM in simple terms. It is also of immeasurable importance in facilitating implementation of the process.

First, let's look at what causes most RCM programs to ultimately result in failure. By examining these issues you will be afforded the opportunity to avoid them.

2.3 Reasons for Failure

Some of the more significant reasons for this lack of success in the past include the following, which are not listed in any specific order.

2.3.1 Loss of in-house control

One of the biggest pitfalls is to farm out the complete analysis to an outside party. While it is acceptable to use outside help for staff augmentation, it is highly recommended to have any outside support work under your direction. You may be thinking . . . "How can they work under my direction if I don't know enough about RCM?" That line of thinking will change as you continue through this book. You will acquire the expertise to bring your own program to fruition. I have seen so many programs fail because an outside team was brought in to do the complete analysis with minimal input from those who really know what is going on in their facility: you and your own people. Just think what happens when subsequent technical questions arise or the database put together by the outside party needs to be changed and the outside party is no longer on-site. Even worse to contemplate is having them come back on-site to make changes at a hefty cost for doing so.

I cannot overemphasize the importance of maintaining in-house control over your own RCM program. I explain how classical RCM can be made so simple that, other than possible staff augmentation working under your direction, you will be able to set up the program, establish the criteria and parameters, maintain authority for the analysis, and make all the requisite decisions for successfully implementing the process at your facility.

2.3.2 An incorrect mix of personnel performing the analysis

I have seen numerous RCM programs result in failure because they did not incorporate the correct mix of knowledge and community buy-in from all applicable in-house parties. If you don't have the consensus buy-in by your maintenance, engineering, and operations personnel, your program will most likely result in failure. If you are fortunate enough to progress toward completing a program but did not get a consensus buy-in up front, what do you think will happen when the first challenge arises questioning the validity for performing or not performing a specific PM task? More than likely, a sense of suspicion will be cast over the entire effort by those stakeholders who were not involved, and confidence in the program will diminish.

What if the RCM analysis was performed only by the in-house engineers, and maintenance and operations people were not given the opportunity to supply their input and wisdom? The left-out parties (with their bruised egos) will probably not embrace the effort, especially if there is a difference of opinion with respect to a decision that was made regarding the consequence of a failure. Remember, plant knowledge is not a monopoly within any one organization. It takes the cumulative knowledge from all associated parties to effect a premier RCM analysis. The RCM effort will invariably lose credibility if any one group does not understand or agree with the basis for specifying a specific preventive maintenance task, or should they ever question after the program has been completed, "Why are we doing this PM task?"

To start out right, I recommend that representatives from the different stakeholder organizations, most commonly engineering, maintenance, and operations, all participate from the beginning. This is true whether you have a large or a small organization. I also highly recommend using craft personnel as temporary members of the RCM team because they can be very insightful, not only for their hands-on knowledge, but also because they will be the ultimate emissaries of the PM program once it is completed. A very important consideration is having the craft understand why they are performing a given task. If your craft personnel understand the bases for which you have established your preventive maintenance program, you will find that they become willing participants and allies instead of adversaries.

2.3.3 Unnecessary and costly administrative burdens

I have seen RCM efforts bogged down for months (while the cost clock is running on fast speed) just deciding which boundary to use or how to prioritize the importance of which system to start with. There have actually been formal probabilistic studies costing many thousands of dollars for the sole purpose of determining which system to analyze first! Just think how that would appeal to your management personnel responsible for paying the bills. As I discuss later, for an average-size facility, the entire RCM process from conception through implementation for all system components should not exceed a few weeks or a few months, at most.

There is no end to the administrative burdens that can be generated. This can include RCM steering groups, focus groups, establishing humongous committees, setting up entirely unnecessary specialized facilitator training programs, and on and on. From many years of experience, I believe that what you really want to know is . . . "How can I establish a premier RCM program without the need for outside "expert" intervention and without all of the unnecessary administrative minutiae?" It is my belief that you have the wherewithal to figure out what committees and steering groups you may need for your organization. Your organization may not want to create an RCM "empire," laden with unnecessary resources.

In later chapters I discuss in detail how to avoid creating administrative nightmares out of the RCM process. I articulate the bigger picture so that, once versed on these principles, you can readily provide the administrative details to the degree that you and your individual organization deem appropriate.

2.3.4 Fundamental RCM concepts not understood

Commencing an RCM analysis without understanding the fundamental concepts can indeed result in failure of the program, but even worse, it can result in a flawed program with a disastrous outcome, even a fatality.

There are numerous real-life examples of disastrous consequences of failure that could easily have been avoided by understanding the concepts I set forth in this book that were never before considered in RCM programs. These concepts are discussed in detail in Chapters 3, 4, and 5.

2.3.5 Confusion determining system functions

Many books and publications are almost exclusively devoted to details about selecting systems and identifying system functions. The customary approach to RCM has been to identify the functions of various subsystems within a larger system. This is one of the first steps in RCM, and unfortunately it is also where the first impact of confusion manifests itself. You might ask yourself, "How do I define all of the system functions? Where do I find these functions? How do I know if I missed a function? What if I did miss a function?" These are all valid questions to be contemplated, and Chapters 3, 4, and 5 explain how to navigate above these problems.

The only reason for identifying system functions and functional failures is the quest to identify the *consequence of failure* as a result of a component failure. In Chapter 3, I introduce a revolutionary new concept regarding the consequence of failure that simplifies the entire RCM process.

2.3.6 Confusion concerning system boundaries and interfaces

Current RCM practices call for defining boundaries and interfaces for each system and subsystem separately. Where does one system end and the other begin? This requires that you define system boundaries, subsystem boundaries, in-system in-

interfaces, in-system out-interfaces, out-system in-interfaces, out-system out-interfaces, system functions, subsystem functions, failures of those functions, consequences of those functional failures, and on and on. Tell me that this isn't confusing! Yet that is the rigor one must go through to implement the archaic approach to RCM. Since system boundaries are completely arbitrary and totally subjective, I have seen this part of the process become a source of much internal bickering and wasted time. Many attempted programs never get beyond this point. Yet this entire part of the process can be eliminated. You will see how to simplify this hurdle and develop a premier RCM program with the same robustness that Nowlan and Heap envisioned in 1978.

What most people do not know is that Nowlan and Heap started with system and subsystem boundaries and functions because, in commercial aviation, they were already there. Few people in industries other than commercial aviation are familiar with Air Transport Association (ATA) codes. Each aircraft, regardless of its manufacturer or type, uses the same coding system. For example, a Boeing 747, a Douglas MD-11, or an Airbus A300 all use the same ATA designations for their aircrafts' air-conditioning systems (ATA 21) or pneumatic systems (ATA 36) or landing gear systems (ATA 32). Nowlan and Heap did not have to develop these system boundaries.

In fact, Nowlan and Heap started at the system level only as a matter of convenience. *It was not a requirement.* It was the component functional failure and its manifestation at the aircraft (or plant) level that was really important to them. What has transpired since then is the interpretation by modern-day RCM practitioners espousing as a requirement the identification of system boundaries, subsystem boundaries, interfaces, and so on.

This has evolved, and continues to evolve, only as the continuation of a misunderstood interpretation. Establishing functions at the system and subsystem level is *not* a required part of the RCM process! In fact, it may even *diminish* the accuracy of the analysis, which may be a revelation to most people. I discuss this further in Chapter 3.

2.3.7 Divergent expectations

There may be an inevitable clash of expectations by your senior management and middle management. As I mentioned earlier, RCM is *not* a PM reduction program. Rather, it is a reliability program. Suppose your senior management has been the recipient of a slick sales presentation extolling the merits of implementing an RCM program and how the results of that program will result in an overall workload reduction. Their first expectation is likely to be determining how many people can be jettisoned. How do you think a less-than-altruistic midlevel maintenance manager or maintenance supervisor would respond to that message? He or she may want to throw a blanket over the RCM program to avoid the possibility of letting people go. It is not a matter of letting people go; it is more an opportunity to reallocate resources so they can be used more efficiently, perhaps through the reduction of any work backlog or overtime needs. I do not subscribe to laying people off as the result of implementing a successful RCM program.

On the other hand, what if an RCM analysis identifies that more maintenance work and more resources will be needed? A less-than-altruistic vice president may want to stifle the effort because of the costs involved. Remember, the results are what they are. If more PMs are needed, it will make the plant more reliable and this will save much more in the long run than a short-term vision of cost savings. In many cases, senior management may depend on the continuation of a lucky streak in that nothing catastrophic has happened yet, so why worry about it? Believe me, when something catastrophic does happen, as it inevitably will, as the result of an unanalyzed failure mode, the associated costs will be magnitudes greater than having had implemented a few critical PMs that could have prevented the catastrophe.

As you can see, if the RCM effort identifies that too much unnecessary work is being done, you run the risk that middle managers may want to scuttle it for fear of losing their people. If the RCM effort identifies that not enough work is being done and critical PMs need to be added, you run the risk that senior management will want to scuttle it for fear of increasing costs. If this mind-set exists in your facility, it should be openly discussed so that neither of these polarizations enters the decision process for attaining a premier RCM program.

2.3.8 Confusion regarding convention

Another reason program implementation attempts have sputtered is because of simple confusion: How do you define a failed valve? Does the valve fail in the open position, or does it fail to close? Which system does a heat exchanger belong to? Is it the system supplying the shell-side cooling medium, or is it the system receiving the benefit of the heat exchanger via the tube side? How do you handle manual valves? Are they governed by a preventive maintenance (PM) strategy or by a corrective maintenance (CM) strategy? To what level does RCM analyze components in an electronic box? Can the functions of identical component types used in different applications be grouped together? *(Absolutely not.)* These issues are all explained in this book.

It should also be noted that most RCM publications, including this book, do not normally include structures. The RCM process is primarily for active functional components unless a review of CM history specifically identifies the existence of a problem with a normally passive component or a structural item. Similarly, manual valves, which are considered passive components, are not normally analyzed except when they are functionally operational in a system, are controlled by operator actions during plant evolutions, or have experienced problems resulting in a history of corrective maintenance. In fact, any CM history identifying specific problems with any passive component or structural item is justification for including it within the preventive maintenance program.

2.3.9 Misunderstanding "hidden" failures and redundancy

These are the most important, yet least understood, concepts. How do you handle "hidden" failures and redundancy? Every RCM book or publication talks about hidden failures, but how do you find them, and what distinguishes them from a run-to-failure component? The challenge is, how do you analyze an entire system that is hidden, such as an emergency safety system? How do you handle hidden failures in hidden systems? In redundant systems? In backup systems? In standby systems? Understanding these differences is a major key to achieving a successful reliability program.

These concepts are all explained in detail in this book. In fact, these concepts are so misunderstood that a recent document on RCM stated that "multiple failures and redundancy are not even considered," and any redundant components were allowed to run to failure with no mention whatsoever of whether that redundancy considered the absence-of-failure indication. That's dangerous.

2.3.10 Misunderstanding run-to-failure

This is also a key concept, and very few people understand it. I explain my "canon law" for run-to-failure in great detail in the following chapters. Wrongly invoking a run-to-failure strategy can have disastrous effects. Most books on RCM state that if a component fails and nothing happens, it is a noncritical, run-to-failure component. This is not only wrong, it is actually dangerous and will inevitably lead to a most unwanted outcome! Likewise, many RCM books state that if there are redundant components they can be classified as run-to-failure components. (Again, dangerously wrong!) Remember, there may not be a preventive maintenance strategy for run-to-failure components, but there is most certainly a corrective maintenance strategy for them. I explain all of this in Chapter 3.

2.3.11 Inappropriate component classifications

Most RCM programs pigeonhole components into either a critical or a noncritical basket. This is another example of what causes a program to flounder. Those categories are too broad. For example, if a failure can result in an immediate effect to the facility, it is usually classified as a critical component. Likewise, if a failure does not result in an immediate effect, but it could, this is also classified as a critical component. Try explaining that to your senior management.

It is also totally imprudent to include a critical component whose failure can result in the immediate shutdown of your plant in the same basket, with the same relative importance, as a purely economic component that will result in only a relatively small dollar cost when it fails. I explain the differences between *critical components, potentially critical components, economically justified components, commitment components, and noncritical run-to-failure components.*

2.3.12 Instruments were not included as part of the RCM analysis

I devote an entire chapter on how to analyze instruments. This is very seldom discussed in any RCM publication.

I am fortunate that my background and experience affords me the opportunity to be intimately familiar with the shortcomings and pitfalls of inappropriately embracing an RCM effort. The reasons identified in this chapter are just a few of them. You will learn how to circumvent these barriers so that when you have finished this book, if you possess just a modicum of intelligence (which I know you do), technical or otherwise, you will be able to implement a successful RCM program at your plant or facility. Neither an engineering degree nor any special training is required for understanding this novel approach to RCM.

Let's look at some of the important concepts that I have mentioned.

Chapter

3

Fundamental RCM Concepts Explained, Some for the Very First Time: The Next Plateau

In this chapter, I introduce to you the fundamental concepts of *Classical RCM Implementation Made Simple*. Once you have mastered these (and I assure you that you will) the subsequent analysis and implementation of an RCM program will become nonproblematic and will fit into your perspective of reliability like a glove.

These RCM concepts, some of which have never before been explained or written about, form the very core of reliability. These are the principles that, once understood, will remove the fuzziness and unveil the mystique that has shrouded RCM. In Chapter 5, you learn how these concepts and principles are used in the RCM implementation and decision logic part of the analysis. Since they form the very core of RCM, understanding them is of paramount importance.

This chapter introduces the totally new concept of a "potentially critical" component that addresses hidden failures. It also introduces another new principle that I call the "canon law" for run-to-failure, which addresses run-to-failure components and corrective maintenance. Another new principle is introduced for delineating what determines a strictly "economic" consequence of failure. You will understand how important these concepts are when you see how they can prevent major catastrophes. After you begin to understand these fundamentals, you will see how relatively simple they are and how they can avert a catastrophe such as the one discussed later in this chapter. The *consequence of failure analysis* (COFA) is also introduced.

It has been my experience that even though the RCM process is a straightforward one, there are many opportunities to easily stray off course and become confused during the process. If you have previously attempted to implement an RCM program on your own, you will understand my point. In baseball parlance, you can wander off from the batter's box and end up in the outfield before you can bat an eye. I have seen many attempts where the individuals responsible for putting an RCM program together are left wandering around in the outfield, unable to get back to home plate. A lot of different pathways need to be brought together in a cohesive manner. If you do not take the logical approach to RCM, which I explain in detail, you may end up expending a great amount of tangential energy without any linear vector to it. Going around in circles has not been an uncommon RCM experience. I have learned that a very regimented but logical mind-set is needed, and, as with any language, the respective alphabet must be mastered first.

The alphabet of RCM lies in understanding the principles and concepts, and that is why I go into elaborate detail to familiarize you with these principles: Once you commence the analysis, the energy you expend will all be linear and none of it will be tangential.

Entities smaller than a nuclear power plant, which includes most entities, cannot afford the unnecessary administrative burdens associated with the current versions of RCM. But the individuals responsible for reliability at these small and midsize facilities still want a robust analysis for their plant. This can be achieved very easily. For the simplified approach to classical RCM, you do not need to set up system boundaries, establish interfaces, and identify functions at the system and subsystem level, which are all elements of the process that are associated with other renditions of RCM programs. That may seem like a revelation, and indeed it is. It should be well understood that the absence of these burdens does not diminish the robustness of the analysis and, as I explain later in this chapter, leaving out that unnecessary work will actually enhance the accuracy and completeness of your RCM program.

Nowlan and Heap's work in 1978 included operational consequences as an economic consideration. Safety was a very specific concern and was seldom invoked unless there was a design flaw or other major design concern. Other than aircraft safety issues, all other operational concerns, including such issues as delayed or canceled flights, were boiled down only to economics.

The RCM program I developed in 1991 addressed operational considerations as being more than just an economic consequence. I made the distinction that operational consequences were either *critical* or *potentially critical*. After all, everything eventually boils down to an economic consideration—even a disaster. However, if you should experience a failure that causes a major environmental impact or some other major unwanted failure consequence, it will be more than merely an economic one for which the penalty is paying a fine. It could result in grounding your fleet of aircraft or having to shut down your plant. It could be the end of your business. A failure consequence can cause such unwanted publicity that calling it an economic issue and framing it as such would do injustice to your facility and would be an inappropriate classification.

To its credit, in 1999 the Society of Automotive Engineers (SAE) also realized this important distinction regarding economic consequences: The differentiation between strictly economic failures and operational-type failures is now a part of the SAE JA1011 RCM Standard.

Also in 1991, I was very specific in the RCM program I developed, segregating hidden failures from evident failures, and I go into great detail, as you will see, on how to identify and manage hidden failures. Again, to its credit, in 1999 the SAE also realized the importance of hidden failures and has specifically included the explicit distinction between hidden and evident failures in its JA1011 RCM Standard. I discuss more about the SAE Standard in Chapter 5. In the next section, we look at the three phases of an RCM program.

3.1 The Three Phases of an RCM-Based Preventive Maintenance Program

Phase 1 consists of *identifying* equipment that is important to plant safety, generation (or production), and asset protection.

Phase 2 consists of *specifying* the requisite PM tasks for the equipment identified in phase 1. These tasks must be both applicable and effective.

Phase 3 consists of properly *executing* the tasks specified in phase 2.

Phase 1 is the genesis of the process. It is where you identify the equipment that must have a preventive maintenance strategy to prevent failure and remain reliable in order to preserve critical equipment functions and minimize any challenges to your plant or facility as a consequence of their failure. This is the equipment population that requires preventive maintenance. This is the equipment population that is deemed important for preserving the "asset reliability" criteria that you establish (which is discussed in Chapter 5).

Once this population is identified, then you can specify, in phase 2, the type of preventive maintenance that should be prescribed. The selection of possible tasks is very large, and newer predictive maintenance (PdM) techniques offer additional cost-effectiveness for accomplishing these tasks. Remember, too, it is not just the tasks that the maintenance department performs; it is the integration of all of the tasks performed by all departments that make up the PM program. The different types of preventive maintenance tasks and how to integrate them is discussed in Chapter 6.

Once you have identified your target population and prescribed the types of preventive maintenance activities that must be performed to maintain their reliability, in phase 3 it is imperative that your work control or work management organization schedules this work at the periodicities specified. What do you think the outcome of your efforts will be, even with a stellar RCM program, if there are deficiencies in scheduling the work, possibly resulting in PM tasks continuously being late or, worse yet, not being accomplished at all?

While phases 1 and 2 are totally under the RCM umbrella, they are also totally dependent on phase 3, which is non-RCM-related and resides under the auspices of your work control and scheduling organization. A breakdown here is like a broken leg on a three-legged stool. While it is not imperative that your work control and scheduling groups participate in the RCM process alongside of engineering, operations, and maintenance, they should not be kept in the dark about the program philosophy, and they should know how it will relate to their scheduling efforts.

I have seen many breakdowns in executing an RCM program because of the difficulties encountered within the work management organization. Depending on your industry and the type of your facility, getting the scheduled work executed properly involves mastering the same sets of problems, although some of the specifics may be slightly different. If your organization is not equipped to deal with the smooth planning, scheduling, and performance of the prescribed work, you may find difficulties in achieving your reliability goals. From personal experience, this is a potential source of weakness that can pose major problems to an otherwise excellent reliability program.

As I mentioned in Chapter 2, RCM is not a PM reduction program. It is a reliability program, and you may end up reducing work or adding work. Reducing scheduled work will offer you the opportunity to reallocate your resources accordingly. However, should the analysis identify that you have not been doing as much preventive maintenance as is required for your plant, additional personnel resources may be needed.

The RCM program I developed and implemented eliminated several thousand unnecessary PM activities, but several hundred new PMs were added to the maintenance program. The added ones were the most important outcome of the analysis. Some typical examples of the added PMs included: identifying protective devices that were not being maintained; identifying rather innocuous components, thought to be unimportant and noncritical, whose failures would have very significant unwanted plant consequences; identifying potential plant shutdown components that had previously escaped inclusion in the PM program; and identifying an entire system that was being maintained but was found to be unneeded and that was subsequently eliminated. Innumerable hidden failures of important components, representing serious potential plant vulnerabilities, were identified and preventive maintenance tasks implemented accordingly. These were just a few of a myriad of very important reliability issues that were found during the analysis.

Another rather interesting outcome of the analysis was the identification of dozens of check valves showing that if they failed while open there would be no consequence of that failure even with additional failures. This revealed that the check valves were not even needed in the plant and that they could be replaced by a spool piece. If they failed while closed, however, there would be an unwanted consequence, so having them in the plant was not only needless, it actually created an unnecessary potential failure mode. Streamlined versions of RCM would not have identified this anomaly.

Let's look at the concepts and principles of RCM in greater detail. I reiterate the techniques for identifying components that must have a preventive maintenance strategy in several different ways. I have designed this chapter as a set of building blocks, beginning with the foundation and sequentially building upward. Let's begin at the foundation, which will help you understand a little more about RCM and how simple it really is. There are three cornerstones to an RCM program.

3.2 The Three Cornerstones of RCM

1. Know when a single-failure analysis is acceptable and when it is not acceptable.//
2. Know how to identify hidden failures.
3. Know when a multiple-failure analysis is required.

These three cornerstones are the bedrock for understanding RCM and I introduce them in only their most simple terms at this time. They are discussed in greater detail throughout this chapter.

1. *Single-failure analysis.* RCM is a single-failure analysis *except* when the single failure is "hidden." When the single failure is hidden, RCM then becomes a multiple-failure analysis.

2. *Hidden failures.* When a component is required to perform its function and the occurrence of the failure is *not* evident to the operating personnel (i.e., the immediate overall operation of the system remains unaffected in either the *normal* or *demand* mode of operation), then the failure is defined as being *hidden*. Addressing hidden failure modes is one of the key aspects for attaining plant reliability.

3. *Multiple-failure analysis.* A multiple-failure analysis is *required* when the occurrence of a single failure is hidden.

Figure 3.1 shows a typical system operational schematic. At first it may appear complex; however, once you realize that you will be analyzing each component individually rather than ana-

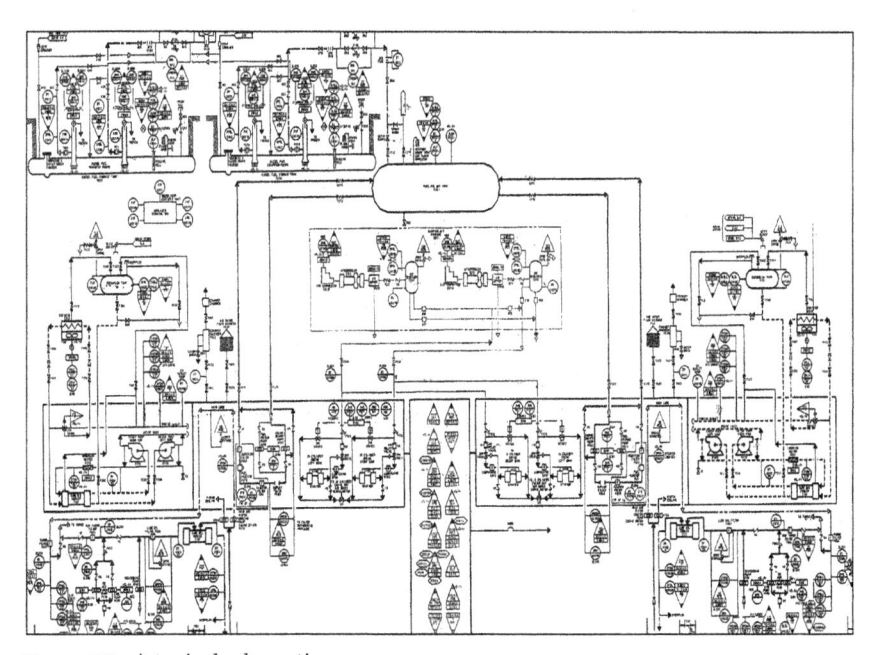

Figure 3.1 A typical schematic.

lyzing the entire system all at once, the intimidation factor is completely dissipated. The illustrations and examples I use throughout this book are simplified excerpts from real-life system schematics and piping and instrumentation drawings (P&IDs). The reason for this is to show how simple it really is when you stay at the component level when analyzing functions. No matter how complicated the plant or system schematic may be, you ultimately simplify it by reducing it to its individual components.

A question often arises about grouping similar-type components together. It is not an accepted practice to group *functions* of similar types of equipment together because the same type of equipment will have very different functions depending on where and how it is incorporated into the plant design. For example, motor type XYZ in one application may have a very different function from the same type of XYZ motor in another application in the plant. However, it is very effective to group *tasks* together for the same type of equipment. The periodicities of those tasks, though, may be very different, again depending on their individual applications in the plant. For example, large motors of a certain horsepower may have similar PM tasks specified, such as thermography, Megger testing, and motor current signature analysis, but the periodicities may be different depending on their cumulative operating times, environment, or other distinct installation or design differences. I discuss more about PM tasks and periodicities in Chapter 6.

3.3 Hidden Failures, Redundancy, and Critical Components

Most RCM books go into a lengthy explanation of the peripheral administrative aspects of the process, such as establishing system boundaries and interfaces, which you have learned are not even necessary to the analysis. They give only a perfunctory or short-shrift discussion to the very essence of the RCM process, which is the understanding of hidden failures, the difference of true redundancy, analyzing a backup or standby function, determining when run-to-failure analysis can be invoked, and when a multiple-failure analysis is required. Since these concepts are so very important, let's see how a single failure, hidden failures, multiple failures, redundancy, and backup functions all fit together by looking at the following examples.

Figure 3.2 illustrates a very basic single-failure analysis. Assume pump ABC provides fuel oil flow to fulfill a critical function for starting an emergency diesel generator. In this illustration, if pump ABC fails, the loss of that function will be evident in two ways. The operation of the pump is monitored continuously by instrumentation in the control room, which will reveal if the pump fails, and second, the inability of the diesel to function when called upon to start will also be evident due to the lack of fuel flow. It is very simple to see that failure of the pump will result in an unwanted consequence to the plant, especially in the event of the loss of on-site power with the alternative diesel power source not being available. This would be a very significant consequence, and a PM task would be required to ensure the reliability of the pump. This is a typical single failure where the failure is evident and causes an immediate unwanted consequence. The component would therefore be classified as a *critical component,* meaning the occurrence of the component failure is evident when it fails and its failure has an immediate unwanted plant consequence, in this case, the loss of a safety feature.

Condition: Pump ABC supplies the function by itself. If pump ABC fails, it is not hidden since there is applicable control room indication.

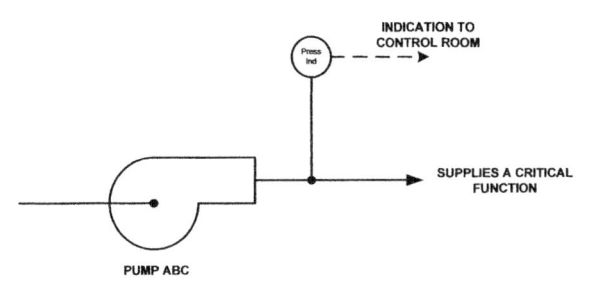

Figure 3.2 Single-failure analysis. A single-failure analysis is acceptable. If pump ABC fails, it will be evident and will have an immediate unwanted consequence of failure.

Now let's look at a similar configuration. In the example shown in Figure 3.3, the designer of the diesel, or the designer of the plant, felt that it was so important for the diesel to start on demand, that he or she incorporated two fuel oil pumps, each one with 100 percent capacity so that either one had the capability to start the diesel. In addition to pump ABC, the designer also added pump DEF to operate simultaneously with pump ABC, but with a separate flow path so the diesel would never fail to start even if one of the pumps should fail. The pressure indication transmitter, which is continuously monitored in the control room, remained downstream of both pumps. Now let's analyze this scenario. Let's assume pump ABC fails. Is the failure evident? The answer is no. It is not evident because there is no pressure indication for each individual pump, and since either pump could supply the required fuel oil flow, pump DEF would continue to supply the required flow and the downstream pressure

Condition: Both pumps are operating and either pump can supply the function by itself. If one pump fails it is hidden since the control room indication is downstream of both pumps.

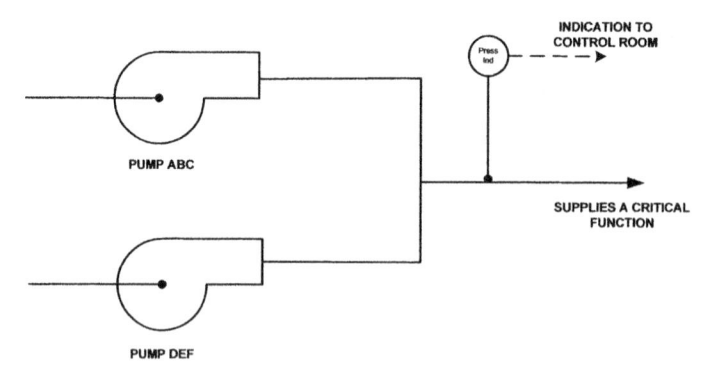

Figure 3.3 Multiple-failure analysis. A single-failure analysis is *not* acceptable and a multiple-failure analysis is required. If pump ABC (or DEF) fails, it will *not* be evident and will not have an immediate unwanted consequence of failure.

transmitter would continue to read the correct pressure in the control room. What has happened here? We have two apparently redundant pumps, and one of them has failed but no one knows it. The failure of pump ABC therefore is hidden, and a plant vulnerability remains undetected. The redundancy feature specifically included by the designer for enhanced safety and reliability has been totally negated. The diesel is now at the vulnerability of a single failure of pump DEF.

Many RCM books and articles and, unfortunately, many "consultants" would consider pump ABC to be a run-to-failure component because, in their thinking, it doesn't matter if it fails since nothing happened when it failed and there is another redundant one. *Wrong!* That is the very trap that has resulted in so many unwanted catastrophic events. In fact, that is a major reason why RCM has been so elusive to implement successfully. Failure of pump ABC is a *hidden failure!* Therefore, a single-failure analysis is *not* acceptable and a multiple-failure analysis is *required*.

To analyze the consequence of the hidden failure of pump ABC requires us to look at what additional failures, in conjunction with the failure of pump ABC, could result in an unwanted consequence to the plant. Since pump ABC's failure is hidden and could remain hidden in a "sleeper cell" mode for some time, what if pump DEF should fail? Then we would have the same situation analyzed in Figure 3.2. An immediate plant effect would occur but only after both pumps had failed. Therefore, pump ABC is a *potentially critical* component. Why a "potentially critical" designation? That is the heart and sole of reliability, and it is discussed in great detail later in this chapter.

In Figure 3.3, since the same situation could occur if pump DEF failed first, it too would remain hidden until pump ABC failed, so they are both considered to be potentially critical components. That is, they have the *potential* to result in a consequence of failure to the plant with the occurrence of an additional multiple failure. The reason there are two pumps is because a single-failure design was *not* acceptable. What if the designer incorporated 100 fuel oil pumps, all in parallel, with no indication of failure for each pump individually? You could have 99 of them fail without knowing it, and you would ultimately be back to single-failure vulnerability.

It creates a false sense of security to assume that just because there are two pumps, you automatically have a guaranteed redundant backup and therefore these are non-critical, run-to-failure components. They are not! Remember, failure of pump ABC (or DEF) by itself will not result in an immediate plant effect. It takes the second, additional failure to manifest the immediate plant effect.

Now that we are beginning to understand a little bit more about hidden failures, let's look at the example in Figure 3.4. In this illustration, note a slight difference. Each pump has its own indicating instrumentation monitored by operating personnel in the control room. In the absence of any regulatory requirement that both pumps must be operable, and in the absence of any economic consideration for the cost of labor or the cost for piece

Condition: Both pumps are operating and either pump can supply the function by itself. If one pump fails it is not hidden since there is control room indication for each individual pump.

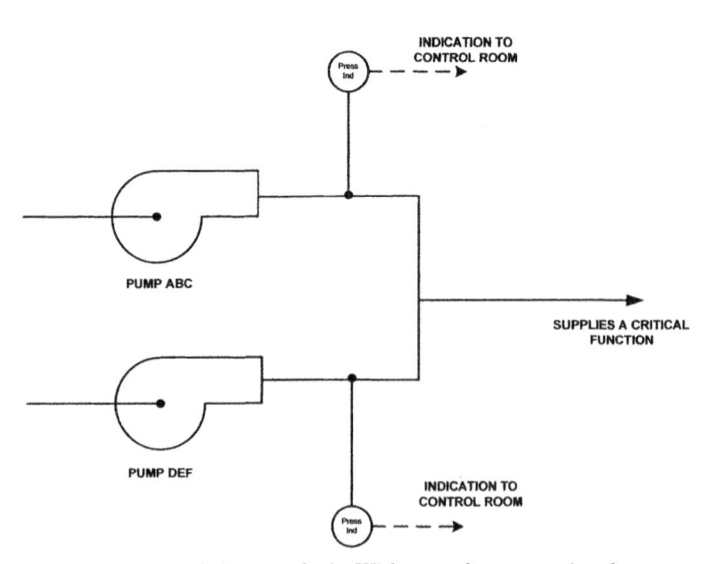

Figure 3.4 Run-to-failure analysis. With no safety, operational, or economic consequence as the result of a single pump failure, each pump is a run-to-failure component since the failure of each one is evident to the control room. Corrective maintenance must be performed in a timely manner when either pump fails.

parts, pump ABC (or DEF) would be run-to-failure components because when one fails there is no immediate effect, but more important, there is an *indication* that it has failed, allowing for corrective maintenance to be performed in a timely manner prior to failure of the other pump. The vulnerability to the plant is no longer undetected by the failure of either pump.

Now let's look at one more scenario. What would happen if pump DEF was the backup or standby pump for normally operating pump ABC and it was available only in the event of a failure of pump ABC? Let's study Figure 3.5 to review this scenario. As you can see, backup pump DEF becomes *critical,* because if it doesn't function when the normally operating pump ABC fails, an unwanted consequence of failure will occur (i.e., the loss of the safety feature). If either pump can act as the backup for the other, then they are both *critical.*

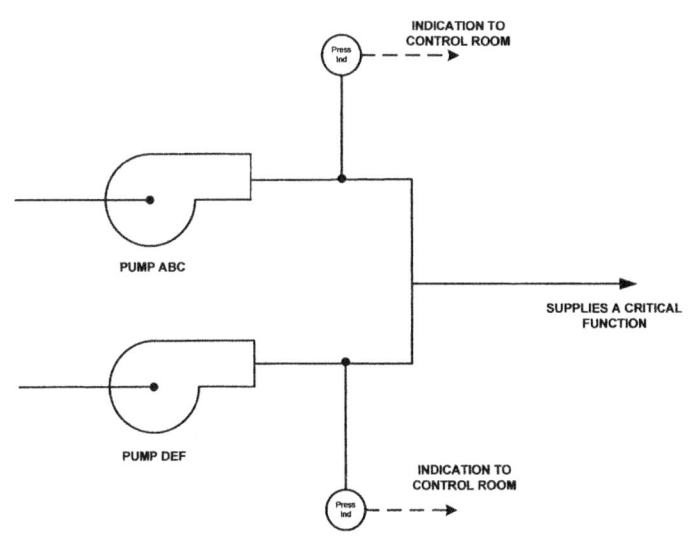

Figure 3.5 Backup function analysis. Pump DEF is a critical component. Although it is not operating, its function is to provide the backup if pump ABC should fail. Since each pump can act as the backup for the other, they are *both* critical components.

The four previous examples were meant to show how subtle differences can result in an outcome ranging from a component being classified as a *critical* component to a *run-to-failure* component. Figure 3.6 shows these differences in one illustration. Incorrectly analyzing a component can have disastrous results, as you will see later in this chapter.

The previous illustrations were real-life examples; now let's look at another real-life scenario as illustrated in Figure 3.7. Throughout the examples and illustrations, I depict the symbol for valves or pumps as part of the illustrations. However, the examples could just as easily have been representative of motors, level switches, safety devices, or any other type of component.

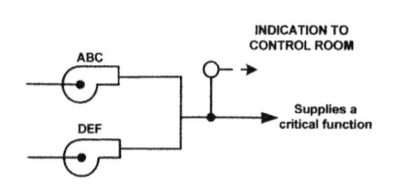

If both pumps are operating but only one pump is required to satisfy the function and the other pump fails, that failure is hidden. Therefore the failed pump is a "<u>potentially critical</u>" component. Since either pump could fail first, they are <u>both</u> "<u>potentially critical</u>."

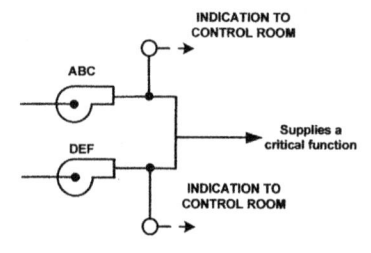

If both pumps are operating but only one pump is required to satisfy the function and the other pump fails, that failure is NOT hidden. In the absence of any safety, operational, or economic consideration, these pumps are "<u>run-to-failure</u>" components.

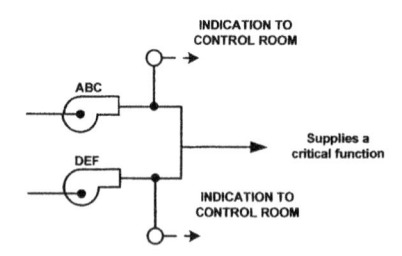

If both pumps are required to be operating together to satisfy the function and either one fails, that is a "<u>critical</u>" component. Since either pump could fail first, they are <u>both</u> "<u>critical</u>."

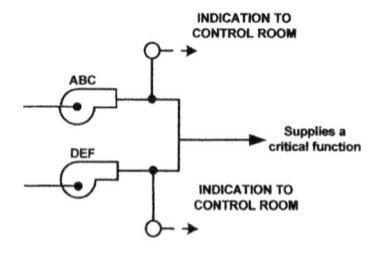

If only one pump is operating and the other is functioning in a backup capacity if the operating pump should fail, the backup pump is a "<u>critical</u>" component. Since either pump can act as the backup, they are <u>both</u> "<u>critical</u>."

Figure 3.6 How apparently similar scenarios can result in totally different outcomes.

Fundamental RCM Concepts Explained 41

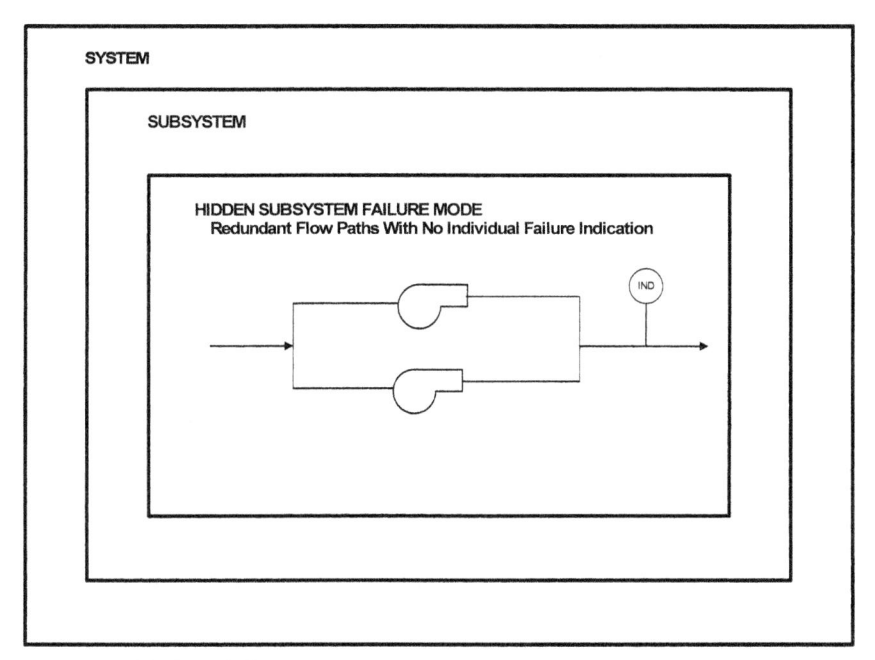

Figure 3.7 A hidden failure.

Does Figure 3.7 look familiar? It should. This is a simplified segment of the schematic for a real-life emergency diesel generator fuel oil system. The fact that only one pump is needed to achieve the function of providing the fuel oil makes this scenario identical to the example in Figure 3.3. The failure of either pump is a hidden failure, because there is no individual instrumentation for monitoring each pump. The function of these pumps is to supply fuel oil for starting the diesel. There are two diesels whose function is to supply AC power to the emergency electrical buses in the event of a local loss of power incident. It is a regulatory requirement that both diesels must be operable. In this case, the diesel will be called upon to start when it receives an undervoltage signal concurrent with the loss of AC power. *When this signal occurs for real, not as a test, is definitely not the time to find out the diesel will not start because both pumps had failed.*

Surprising as it may seem, a primary regulatory operability requirement of the diesel was satisfied by a monthly operability test of the entire unit. As you have now seen and learned, that is not a sufficient test.

Hidden failures are often (but not always) failures of one or more components in a parallel design with no indication of failure for each *individual* component. In this real-life illustration, note that the only pressure indication transmitter is downstream of both pumps. One of the two components could fail, but since each one by itself can satisfy 100 percent of the function and supply the required fuel supply at design pressure, only when the second one fails (i.e., multiple failure) will the total functional failure of the diesel become evident; hence the failure of the first component is *potentially critical*. Since either pump can fail first, both are potentially critical.

To summarize this scenario . . . if one pump failed for whatever reason, the other pump would provide the function and still indicate the acceptable pressure in the control room. The monthly start test of the diesel would be performed each month, and it would continue to be successful. However, it is only a matter of time before the second pump will fail, and then the function of supplying fuel oil will be lost, allowing one of the emergency diesel units to succumb to a functional failure. Since Murphy's Law will ultimately prevail if you allow it enough time, what do you suppose would happen if that emergency diesel were called upon to function in a real-life situation during the interval between the monthly testing schedule . . . and it failed to start? What if the same scenario occurred to both emergency diesels? It could be catastrophic! It is exactly failures such as this that cause the most severe, unwanted consequences to a facility. *Addressing potentially critical failure modes is therefore a key aspect for successfully achieving plant reliability.*

How could this have been avoided? Very simply. At the component functional level you would have identified that the single failure of one pump was hidden. Then you would have known that a multiple-failure analysis was necessary, and you would have identified that if an additional failure of the other pump should occur, you would be faced with a total loss of diesel-starting capability. Therefore, a PM task would be required to address these potentially critical components to prevent their failure.

You would also have seen that even though there were two redundant pumps, that, in itself, did not qualify these components to be noncritical, run-to-failure components devoid of any preventive maintenance strategy. A failure-finding task at the *component* level (not at the system or subsystem level) would be required, at a minimum.

Most people would automatically jump to the conclusion that some kind of a design change was necessary to avoid this potential problem. Not true. In fact, the overwhelming majority of these scenarios can be very easily resolved with an additional PM activity added to your preventive maintenance program. The need for a design change is the exception rather than the rule. The fix for the vulnerability concern in Figure 3.7 was not an expensive modification. It was as simple as adding a sign-off step to an existing operations procedure requiring the operator to "listen" for *both* pumps to be running to verify their simultaneous operation. Chapter 6 discusses in great detail the logic for defining PM tasks and determining when a design change may be required.

Remember, this was just a typical real-life scenario. The same fundamental issues are applicable to an enormous number of similar situations in your facility. In this instance, the RCM analysis identified that pumps ABC and DEF were both potentially critical because either one could fail and not be evident, and therefore both required a preventive maintenance strategy to ensure the reliability of the diesel. Can you imagine if these components were identified as run-to-failure (RTF) components? As noncritical? One of them could have been in its failed sleeper cell mode for a long time just waiting for the other one to fail, in which case a very unwanted consequence would occur to the plant (i.e., no AC power to some of the emergency buses during an emergency). As I mentioned, that is not the time to find out that neither pump will function to provide the starting capability for the diesel. The design objective of having two pumps for redundancy would have been totally negated. The worst-nightmare scenario would happen if this same common-mode hidden failure occurred to both emergency diesels.

Let's look at another typical scenario illustrating hidden failure modes. In Figure 3.8, two pumps are running simultane-

44 Chapter Three

ously. There is a discharge check valve for each one to prevent backflow in the event of failure of either pump. What if check valve C fails open with both pumps running? Nothing. Nothing will happen if the check valve fails open, but the failed-open check valve is not evident, either. Therefore a multiple-failure analysis is required to see what consequence could occur if some

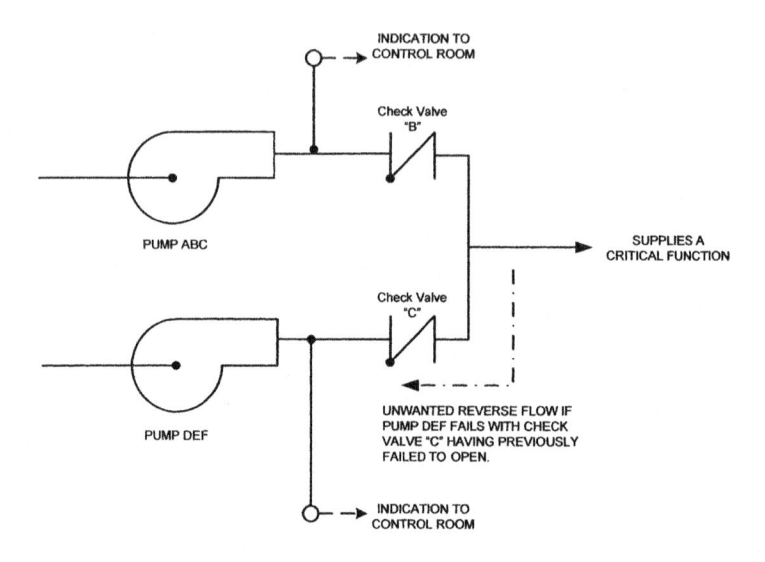

Plant Condition:

- During normal plant operation, pumps ABC and DEF are both running and supplying a critical function.
- Check valve "C" fails in the OPEN position due to hinge pin wear.

Analysis Considerations:

- The failure of check valve "C" in the OPEN position is a "hidden" failure since there is no indication of failure and there is no operational consequence to the plant or facility.
- However, with an additional failure, such as failure of pump DEF, a plant operational consequence will occur due to the diversion of flow from pump ABC through the failed OPEN check valve "C."

Analysis Result:

The effect of check valve "C" failing OPEN during normal plant operation is considered to be **"Potentially Critical"** because its failure is "hidden" and a plant or facility consequence will not occur <u>until</u> the second failure (pump DEF) occurs. Since this same scenario can happen if check valve "B" fails first and then pump ABC fails, it too is considered to be **"Potentially Critical."**

What about the pumps? If both pumps are operating but only one pump is required to satisfy the function and the other pump fails, that failure is NOT hidden. *In the absence of any safety, operational, or economic consideration,* these pumps are **"run-to-failure"** components. The failed pump must have *corrective maintenance* performed in a *timely* manner before the other pump fails.

Figure 3.8 A hidden failure.

other component should fail while check valve C remains in the undetected failed open position.

What happens if pump DEF should now fail? With check valve C being unable to close, a reverse-flow condition from pump ABC would occur, diverting the design flow away from the critical function it was supplying. Therefore, check valve C is potentially critical, and since either check valve can fail first, both discharge check valves are classified as being potentially critical. What about the pumps themselves? In the absence of any safety, operational, or economic concern, and assuming that either pump can supply the necessary flow by itself, both pumps are considered RTF because there is an indication of failure for each one and there is no consequence as a result of the failure of either one. As you will learn later in this chapter, *corrective* maintenance on the failed pump must take place in a timely manner.

3.4 Testing Hidden Systems

In the example in Figure 3.7, you actually have a hidden failure within a hidden subsystem within a hidden system. The entire system is the diesel, which is a hidden standby system. The subsystem is the fuel oil start system, which is a hidden subsystem of the diesel, and the hidden failure mode within the subsystem is the fuel oil pump. As you can see, when analyzing functions at the component level, it clearly illuminates this potentially critical consequence. If you started by identifying functions at the system level, the consequence of failure of the fuel oil pumps would not be as readily visible, and, in fact, unless you have some RCM analysis experience, you may even miss it completely.

Some of you may be thinking, "The entire diesel is a hidden function, so how can I analyze it?" To make the analysis meaningful, *any normally nonoperational system, such as fire protection, emergency or auxiliary power, emergency feedwater, automatic shutdown features, or any other emergency standby systems that do not normally operate, must be analyzed in its demand mode of operation just as though that demand mode were its normal operating mode.*

Equally important is to realize that even a PM task that performs a broad overall test of the entire hidden system, which entails pushing a start button and watching an output light come on, would still not be sufficient to identify the preceding potentially critical scenario. In this uncomplicated, although real-life example, an individual PM task to test each pump individually and verify its operability would be required.

Even today (but hopefully not for long), it is an accepted practice to test an entire hidden system once per month or once per quarter and validate it as being 100 percent reliable and ready for operation in the event of an emergency demand. This is called a *failure-finding task,* and we discuss more about the different types of preventive maintenance tasks in Chapter 6. A failure-finding task is a valid type of preventive maintenance strategy, but it should not be a substitute for internally analyzing the entire hidden system. Even though the hidden system may be tested once per month, once per quarter, or whatever the interval, that test alone will not reveal the potentially critical sleeper cell modes of failure at the component level.

3.5 The Missing Link: Potentially Critical Components

I have already introduced you to the term *potentially critical*. This was to acquaint you with the importance of this concept. I created the concept of potentially critical components because there has never before been an understandable component classification for these hidden failures. It was a void in the process and a source of much confusion. Indeed, it was a missing link of RCM. The "dots" may have been there, and hidden failures may have been obliquely noted, but the customary RCM classifications included only "critical" and "noncritical" component designations. Hidden failures fell through the cracks and were not *specifically* identified since they don't fall into either the critical or the noncritical category. To be a critical component, the failure must be evident and have an immediate plant effect. With a hidden failure, that is not the case. Therefore, since the failure of pump ABC (or DEF) in Figure 3.7 did not have an immediate effect, it would most likely have been classified as noncritical and run to failure, which is definitely incorrect.

Much of the fuzziness of RCM has come about by trying to justify that a component whose failure has no effect on the plant still needs to have a preventive maintenance strategy. If a hidden failure was included in the critical classification, it then had to be explained that a critical component has an immediate effect, but a critical component may also not have an immediate effect if the failure is hidden. This was so confusing that the usual strategy invoked (incorrectly, I might add) was a run-to-failure strategy and the component was classified as noncritical. This was, and still is, one of the major stumbling blocks with RCM, and that is precisely why I have given hidden failures, which could *potentially* result in an unwanted plant effect, their own designation. The *potentially critical* designation provides the conceptual clarity that was missing.

This aspect is virtually totally misunderstood by even some of the most astute engineering types in the industry, even those within the nuclear industry (not that nuclear types have any monopoly on brainpower). That is why the concept of potentially critical components is so important. In my opinion . . . *finding hidden failures and addressing them accordingly, especially in your nonnormally operated hidden safety systems, will afford the greatest opportunity to avoid a catastrophic event.* Later in this chapter, in the "Anatomy of a Disaster," you will see just how simple it is to overlook this aspect without understanding the concept of potentially critical components.

A *potentially critical* component is one whose immediate failure is *not evident* and is *not immediately* critical but has the potential to become critical, either with a duration of time, in and of itself, or with an additional failure or initiating event, at which time the consequence of the failure may unfortunately become quite evident (and critical). The potential to become critical can occur not just with an additional component failure, but also with an additional initiating event, or even with an additional routine plant evolution (turning on another system, turning on a switch, shutting off a pump, etc.).

Oftentimes a potentially critical component may not rear its ugly head until some other initiating event occurs, such as a fire, a steam-line break, a pipe rupture, or a loss of power. I have even seen where a hidden failure was lying in wait, unknown to everyone, until a new operating evolution occurred, such as switching to a rarely used alternate pump or testing a newly modified plant design. Those are not the times to find out, firsthand, that you have had existing plant vulnerabilities that were unknown. This has happened in the nuclear industry and it can happen anywhere in any industry.

Potentially critical components can be thought of as sleeper cells, lying in wait and ready to bring havoc upon your plant or facility. They are failed components that are not evident, and no one knows they have failed. They are components that will no longer function, but you don't and won't know about their loss of function and potential consequence of failure until an additional failure, initiating event, or evolution occurs, causing the sleeper cell to manifest itself. *That is a vulnerability that must be avoided!*

The concept of a potentially critical component is totally different from, and should not be confused with, the *potential failure* of a given component. The *potential failure* of a component refers to an impending precursor, or cause, of a component failure, such as monitoring a bearing for vibration when it has been making noises or monitoring a motor for excessive temperatures when it appeared to be running hot, for example. A potentially critical component refers to the potential consequence of failure to the plant, after the hidden failure of the component has already, albeit unknowingly, occurred.

Note the similarities between *critical* and *potentially critical* components. The only difference is that critical failures manifest themselves immediately, whereas the failure of potentially critical components is hidden and will not manifest itself until a second multiple failure or initiating event occurs or a certain time duration occurs. The majority (approximately 98 percent) of potentially critical components will fail because of (1) the effects of *multiple failures or initiating events* versus (2) the effects of *time* duration (approximately 2 percent). The reason I include the aspect of time duration is for the analysis to be absolutely complete, thorough, and accurate, so that nothing escapes the analysis or is inadvertently omitted.

1. *Potentially critical components as a result of multiple failures or initiating events.* When two (or more) components (valves, pumps, motors, etc.) operate to supply a function that each can fulfill individually, and there is no indication of failure for each component individually, then a failure of one of the components will be hidden (*there will be no indication the component has failed*) and the failure will not result in an immediate plant effect. However, if a second component should fail or if some other initiating event or plant evolution takes place that would otherwise rely on the failed component, then a plant-affecting consequence would occur. Hence, the component is considered to be *potentially* critical. Another typical example is the pump discharge check valve shown in Figure 3.8. If two pumps are normally operating at the same time, a failure of the discharge check valve in the open position will be hidden. Only when the associated pump fails will the unwanted reverse-flow path through the failed open check valve become evident.

2. *Potentially critical components as a result of Time.* Here's a typical example of being potentially critical due to time. If one panel of a multipaneled circulating water traveling screen that filters out seaweed and other ocean debris were to fail or become damaged, there would be no indication that the panel had failed, nor would there be an immediate effect. However, over a duration of time, the failure of one of the screens' panels, *in and of itself,* can eventually cause clogging of the heat exchangers by failing to filter debris, which will ultimately result in a plant effect. Note that the traveling screen never fails completely or immediately and that you do not need to have a second additional failure for this consequence to occur. Here's another example of classifying a component as being potentially critical because of time duration: Suppose a large tank had a small pinhole leak that went undetected. This would not be evident immediately, nor would it require a second (multiple) failure to manifest itself. However, over a given duration of time, in and of itself, a plant consequence may occur given that the inventory of the tank represents an important function. These are vulnerabilities that you will not know about until they have already resulted in a plant consequence, which is not the time to find out about them. When that happens, it is then too late to take preventive actions.

You might wonder how prevalent hidden failures are. They are extremely prevalent. Just a few typical but very important examples include main turbine overspeed components, critical check valves, emergency diesel generators, safety shutdown components, protective devices, fire protection systems, emergency pumps, and critical motors. *They are found everywhere, and it is the belief of this author that correctly identifying potentially critical components affords perhaps the greatest degree of reliability protection you can provide to your plant or facility.* To reiterate a phrase from Chapter 1, it is not the obvious that causes the most unwanted problems, it is the nonobvious!

How important is this concept? *Very.* There are numerous examples in industry whereby a designer intentionally built in multiple redundancy to ensure reliable system operation. Unfortunately, if the redundancy has no way of manifesting itself when it fails, the plant is vulnerable to an unwanted consequence, which can occur with an additional single failure.

Now let's review what we have discussed thus far. A *critical* component is one whose failure must be avoided, either by a preventive maintenance strategy or through redesign. The occurrence of the single failure is immediately evident, either by its indication of failure or by the unwanted consequence it will cause as a result of its failure. Depending on the asset reliability criteria that you establish (which is discussed in Chapter 5), even a failure that does not have an immediate physical consequence when it fails could have an immediate unwanted regulatory, legal, or customer-related consequence when it fails, so it is still considered to be immediately critical.

A *potentially critical* component is one whose immediate failure is not evident in any manner, but with additional failures or initiating events or over time it could become critical, so it, too, must be avoided, either by a preventive maintenance strategy or through redesign. That leaves us with *commitment, economic,* and *run-to-failure* components.

3.6 Commitment Components

I include commitment components as part of the preventive maintenance strategy to ensure that nothing is overlooked in the analysis. Oftentimes your facility may have certain regulatory, environmental, Occupational, Safety, and Health Administration (OSHA), or insurance commitments that must be maintained, thereby requiring a preventive maintenance strategy to preclude a component from failing and causing a commitment to be missed or possibly an infraction of the commitment. Some typical examples of commitments governing certain components are as follows: insurance commitments required for major pieces of equipment, state code commitments required for pressure vessels, Environmental Protection Agency (EPA) commitments related to an environmental impact or a fluid or gaseous release, OSHA commitments required for personnel safety, and federal regulatory agency commitments such as FAA or NRC information notices and bulletins.

Identifying commitment components has an additional positive side to it in that it affords you an opportunity to challenge the validity of having that commitment imposed on your plant. *Usually,* a commitment component is also associated with the component being classified as either critical or potentially critical because of its importance. However, on occasion, you may find a certain component governed by some type of commitment whereby the component was analyzed to be a run-to-failure component. This has occurred within the nuclear industry, so it is not unlikely that it can happen in your industry.

To take this even one step further, there have been entire systems composed of dozens of components that were required to be maintained by preventive maintenance because of commitments whereby the RCM analysis identified the entire system to be run-to-failure. As a result, an entire system was able to be deleted from regulatory required space. I am not suggesting that you automatically delete a commitment if the analysis leads you in that direction *until* you have formally justified it and received approval to delete the commitment from the appropriate agency requiring it in the first place.

3.7 Economic Components

An economic component is one that I refer to as having an *economic consequence only*. Failures of economic components have no effect on plant safety or operability. Economic failures will result only in labor and/or parts replacements costs.

Note: If a failure occurs to a component with an economic consequence but the failure *also* results in an effect on plant safety, operation, or production, it would be more than merely an economic consideration. It would be captured as either a critical or potentially critical consequence of failure, and the component classification would default to that most limiting classification. For example, the failure of a major piece of equipment is obviously an economic concern, but even more important, failure of that equipment will most likely also result in an unwanted consequence such as a plant shutdown, a safety condition, or some other operational concern.

The reason I created this distinct economic classification is because of the prioritization of its relative importance as well as the prioritization for performing work. As I mentioned previously, most RCM programs have either a critical or a noncritical component designation. Therefore, the failure of a small $100 pump, which one might consider an economic concern, would be thrown into the same critical component basket with a component whose failure could shut down your facility. That not only doesn't make sense, it actually dilutes the importance of critical components.

Another reason I maintain an economic consideration as a separate and distinct classification is because it will invariably happen that at certain times your planned workload may not allow for all of the PM tasks to be accomplished when they are scheduled, and some of them may need to be deferred. Of course, this should be the exception and not the rule. However, when this occurs, some sort of priority needs to be placed on the tasks to identify which ones can be deferred. Obviously an economic task associated with an economic component that has no safety or operational consequence of failure would be deferred before a critical task associated with a critical component.

3.8 The "Canon Law": Run-to-Failure Components

The concept of run-to-failure (RTF) is widely misunderstood, perhaps even more misunderstood than hidden failures. I call RTF *the misunderstood orphan of reliability.* It is a given fact that most people, engineers included, will provide the automatic response that if a component fails and nothing happens, it is a run-to-failure component. As you now know, that is totally wrong. Another very prevalent, but totally incorrect, assumption is that having redundant components or redundant systems automatically means the component or system is run-to-failure. Again, totally wrong!

The run-to-failure definitions that exist today do not adequately address the true meaning of RTF, whether they are found in an RCM publication, a regulatory publication, or in fact, any publication. The standard definition for a run-to-failure component usually reads something like this: "The component is allowed to fail without the requirement for any type of preventive maintenance," or "Run-to-failure is a policy that permits a failure to occur without any attempt to prevent it." These definitions are far too shallow to prevent the mismanagement of this very important concept. The time has come for a very precise and prescriptive definition for identifying when a component can be classified as run-to-failure. I have termed this the "canon law" for run-to-failure, and I define it as follows:

> A run-to-failure component is designated as such solely because it is understood to have no safety, operational, commitment, or economic consequence as the result of a single failure. Also, the occurrence of the failure must be evident to operations *personnel.*
>
> As a result, there is *no proactive preventive maintenance strategy* to prevent failure. However, once failed, an RTF-designated component *does have a proactive corrective maintenance strategy* commensurate with all other components based on the plant conditions at that time.

My canon law is very specific. Traditionally, run-to-failure in its most basic definition means that "PMs are not required *prior* to failure." There is no mention that corrective maintenance is "required in a *timely* manner *after* failure." However, that is only part of the RTF story. There are several qualifiers before a component can be classified as run-to-failure.

Run-to-failure components are *understood* to have *no safety, operational, commitment,* or *economic* consequences as the result of a single failure. Also the occurrence of failure must be *evident* to operations personnel.

RTF components have been mistakenly designated as unimportant because they have no significant consequence as the result of a single failure. However, after failure, the component is still required to be restored to an operable status via *corrective maintenance* in a *timely* manner. *RTF does not imply that a component is unimportant!* It is just that some components must have a *preventive maintenance* strategy, and RTF components do not. However, *all* components, even RTF components, are important to reliability and must have an *equivalent corrective maintenance* strategy commensurate with all other components, and prioritized accordingly, based on the plant conditions at that time. You will see how this all fits together later in this chapter.

RTF components are designated as such due to the failure being evident and having no significant consequence as the result of a single failure. If it does not matter whether a failed component is restored to an operable status in a timely manner, one could question why that component was even installed in the plant. Similarly, if the failure remains forever hidden and no one ever knows about it and it doesn't matter how many additional multiple failures occurred, one could also question why that component was even installed in the plant. Figure 3.9a and b summarizes the RTF philosophy.

As I delineated in the first paragraph of this section, it is an absolute fact, born out by my own experience as well, that most people will provide the automatic response that if a component fails and nothing happens, it is an RTF component. The other prevalent, but totally incorrect assumption is that having a redundant component (or even a redundant system) automati-

---

A run-to-failure component is designated as such solely because it is understood to have no safety, operational, commitment, or economic consequence as the result of a single failure. Also, the occurrence of the failure must be evident to operations personnel.

As a result, there is *no proactive preventive maintenance strategy* to prevent failure. However, once failed, an RTF-designated component *does have a proactive corrective maintenance strategy* commensurate with all other components based on the plant conditions at that time.

—Neil Bloom

Figure 3.9(a) The canon law of run-to-failure.

cally means the component (or system) is run-to-failure. *These assumptions are recipes for disaster!* That is the very reason I give so much significance to the concepts of "potentially critical" components, "hidden failures," and the "canon law" for run-to-failure.

Another major misconception in regard to a run-to-failure component is the thought that fixing an RTF component when it is broken is either optional or has no fundamental consideration for a timely repair. This is absolutely incorrect! As I state in my canon law, a run-to-failure component has no proactive preventive maintenance strategy, *but it does* have a proactive corrective maintenance strategy commensurate with all other components (i.e., critical, potentially critical, economic, or commitment) depending on the plant conditions at that time. If an RTF com-

---

Question:

Are run-to-failure components important?

Answer:

*Yes!**

RTF components do not require preventive maintenance *prior* to failure, but then corrective maintenance is required in a timely manner *after* failure. Just because a component may be designated as RTF, *does not* imply that the component is unimportant and that it does not need to be included in a prioritization plan for being fixed. The RTF component is still required to be restored to an operable status via corrective maintenance in a *timely* manner!

*RTF Thought Process:

RTF components are designated as such due to the failure being evident *and* having no safety, operational, or economic consequence as the result of a single failure. If it did not matter whether or not a failed component was restored to an operable status in a timely manner, one would question why that specific component was even installed in the plant.

RTF does not imply that a component is unimportant!

Figure 3.9(b) The canon law of run-to-failure (*continued*).

ponent is indeed of low relative significance, its place on the corrective maintenance hierarchy will reflect that.

Unfortunately, RTF has become the orphan in the picture of reliability. Time and again, engineers and senior management embrace the belief that RTF components are like secondhand junk cars, not worthy of worrying about either before or after they fail. That is totally misguided thinking. That line of reasoning is tantamount to having a flat tire, putting on the spare, and throwing the flat, with the nail embedded, back into the trunk and never worrying about it again. I have even seen some facilities prepare formal documents specifically stating that RTF components are not important to reliability.

One reason for this misguided logic is that preventive maintenance historically, and RCM specifically, has keyed in on only critical components to the detriment of all others. Another possible explanation is the run-to-failure terminology itself. It has somewhat of an ominous impression. I can remember several occasions when I had to use the choice of words that a specific component was "governed by corrective maintenance" just to avoid using the RTF terminology, because the receivers of the information in the conversation were not sufficiently astute to accept an RTF component as being anything other than totally irrelevant. Hopefully, one day, industry on a universal level will come to realize that RTF components are quite important. That is why I go into significant detail in this chapter explaining the different concepts and the different component classifications and why all of them are important.

Another very important principle to keep in mind is that you cannot assume that two components will fail simultaneously at exactly the same time. If you made this assumption, *everything would be critical.* The RCM analysis assumes that only one component fails at any given time. Don't confuse this with a hidden failure that occurs *prior* to a second multiple failure, because those failures did not occur at the same exact time. The FAA and the NRC also recognize that you must assume only one failure occurs at any given time. For critical equipment, these agencies have very specific allowable time envelopes, of a few hours to a few days, to repair a failed component or take other compensatory measures to avoid the risk of an additional failure; otherwise, an aircraft is not allowed to fly and a nuclear plant must be downpowered or placed completely into a cold shutdown condition if corrective measures cannot be taken in time. These requirements are regulatory requirements referred to as the Minimum Equipment List (MEL) in airline terminology, and the Technical Specifications (Tech Specs) in the nuclear power industry.

I introduced the concept of RTF in early 1991 at the nuclear facility where I was working. At that time, the very phrase of *run-to-failure* was anathema within the nuclear industry. Nuclear power was very far behind the airlines in understanding equipment reliability. It was not until many years later that governmental publications governing nuclear power first acknowledged the acceptability of the run-to-failure concept. Even today, many plants are struggling to get a firm handle on identifying exactly what equipment is important to reliability and how to define that population with respect to a preventive maintenance strategy. That statement segues into the following section.

3.9 The Integration of Preventive and Corrective Maintenance and the Distinction Between Potentially Critical and Run-to-Failure Components

A total proactive maintenance plan integrates corrective maintenance as well as preventive maintenance into its strategy. At first, this statement may seem startling, but once you think about it, it becomes quite clear. Let me explain. The *ultimate* objective of preventive maintenance is to prevent a consequence of failure at the plant level. Preventive maintenance tasks are specified to prevent component failures that have either an *immediate* unwanted consequence of failure or the *potential* for an unwanted consequence of failure at the plant level.

Other times, we have seen where it is acceptable to allow a component to run to failure. As we have learned, run-to-failure components, by definition, have no immediate effect on the plant when they fail and that therefore preventive maintenance is not required. However, as stated in our "canon law," the failure *must* be evident. It just means that it is more prudent not to perform preventive maintenance and not to expend unnecessary resources to prevent their failure. *However,* we have also learned that once a run-to-failure component has failed, it *must* be corrected in a timely manner via corrective maintenance. Corrective action must be taken in a timely manner to eliminate or reduce the vulnerability of a plant consequence should another component fail while the RTF component is in its failed state.

My canon law further states that RTF components are still important for the operation of the facility. You don't need to have a proactive preventive maintenance strategy to prevent their failure but you are required to have a proactive corrective maintenance strategy to fix them once they have failed. If you do not impose a proactive corrective maintenance strategy, you run the real risk of an unwanted consequence at the plant level with an additional failure. *Remember, the only difference between a potentially critical component and a run-to-failure component is that the potentially critical component failure is hidden. If its failure were not hidden, it would in all likelihood be a run-to-failure component.* Putting it into perspective, both potentially critical components and run-to-failure components pose a vulnerability to the plant should additional failures occur. *The difference is one failure is hidden and the other is not.*

Preventive maintenance is a strategy to prevent component failures before they occur. Corrective maintenance is a strategy to fix components once they have already failed. These two entities are performed integrally to prevent a failure consequence at the plant level (or whatever other parameter you may select based on the criteria you establish as part of your asset reliability strategy). After a component has failed, whether it was governed by a preventive maintenance strategy or an RTF strategy, it is prioritized for corrective maintenance with an *equivalent relative importance* based on the plant conditions at that time. This requires a decision process, usually by operations and engineering, that considers all pertinent factors (e.g., what other equipment is out of service) in attempting to prevent any possibility of a failure consequence at the plant level. Now it should become much easier to envision how preventive and corrective maintenance coexist to achieve the desired reliability outcome. Refer to Figure 3.10.

Prior to failure:	Those components with prescriptive PM tasks specified (*a proactive PM strategy*)	Those components with *no* prescriptive PM tasks specified (*RTF components with no proactive PM strategy*)
After failure:	*All* components (those with prescribed PM tasks and RTF components) are prioritized for a *proactive corrective maintenance strategy* with an equivalent relative importance based on the plant conditions at that time. This requires a decision process, usually by operations and engineering, that considers all pertinent factors, such as what other equipment is out of service, for example, in attempting to prevent any possibility of a failure consequence at the plant level.	

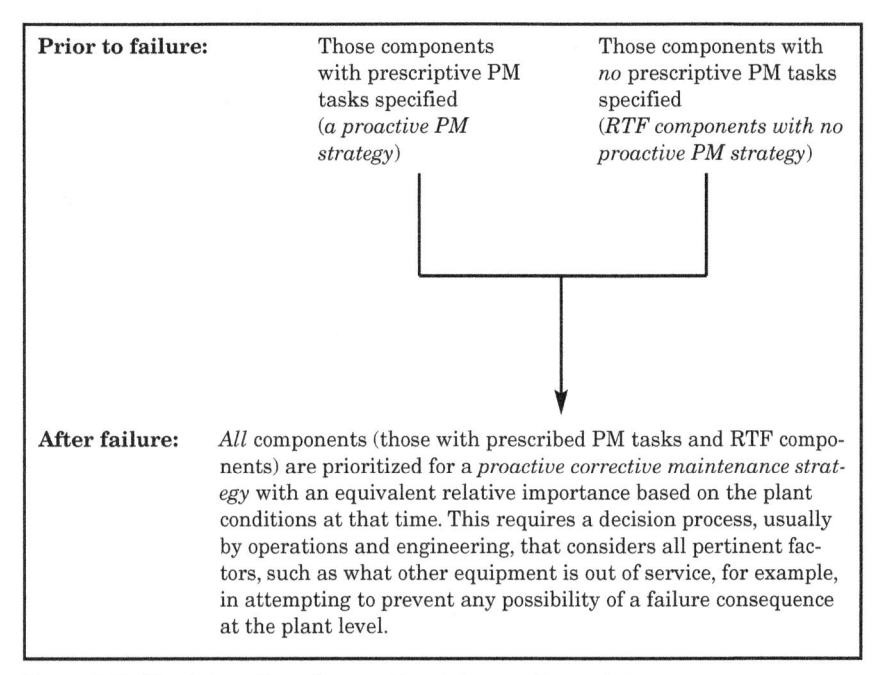

Figure 3.10 The integration of preventive and corrective maintenance.

The traditional vision of preventive maintenance, which I think of as the smaller picture of preventive maintenance, is to prevent failures at the component level prior to the component failure resulting in an unwanted plant consequence. However, I prefer to think in terms of the bigger picture of preventive maintenance, which is to prevent an unwanted consequence of failure not only at the component level but directly at the plant level as well. To do so includes addressing and prioritizing corrective maintenance within a total proactive maintenance strategy, and run-to-failure components are an integral part of this bigger picture.

3.9.1 An RTF CM versus a Critical CM: Which Takes Priority for Getting Worked First?

This is another way of explaining just how important corrective maintenance is in regard to the total proactive preventive maintenance strategy. Let's look at the following quiz.

There are two scenarios:

- *Scenario 1.* Two redundant pumps, A and B, are operating simultaneously. Either pump is capable of supplying the necessary flow to satisfy the function. There is an indication of failure for each individual pump. The failure of either pump will not result in an unwanted consequence of failure, because the other pump can provide the necessary flow. Also, there is no economic concern as the result of failure. Therefore pumps A and B are both RTF. However, if both pumps should fail, an unwanted plant consequence would occur.

    Assume the following:

    1. Pump A has failed previously and a corrective maintenance order (CM) has been scheduled to replace it.
    2. To add a bit of additional realism to this quiz, assume that a member of the housekeeping crew noticed a small puddle of oil accumulating under pump B during the midnight shift and reported it to the maintenance supervisor on duty. A CM was written to fix the oil leak on pump B.

- *Scenario 2.* Only one pump, C, supplies the necessary flow to satisfy the function. It is a critical pump, because its failure will result in an unwanted plant consequence. Pump C just had a predictive maintenance task (PdM) scheduled to sample the oil, and the results of the oil sample showed evidence of an incipient bearing failure.

    Assume the following:

    1. Pump C has a CM scheduled to replace the bearings as a result of the oil sample findings.

- *Scheduled activities:*

    In scenario 1, a CM is scheduled to replace pump A and another CM is scheduled to fix the oil leak of pump B.

    In scenario 2, a CM is scheduled to replace the bearings in pump C.

- *Question.* Which CM takes priority? Is it the CMs on the run-to-failure, redundant pumps A and B or the CM on the critical pump C? The answer can be found at the bottom of Figure 3.11.

Fundamental RCM Concepts Explained 61

SCENARIO (1)

PUMP(A) — INDICATION TO CONTROL ROOM
PUMP(B) — SUPPLIES A CRITICAL FUNCTION
          INDICATION TO CONTROL ROOM

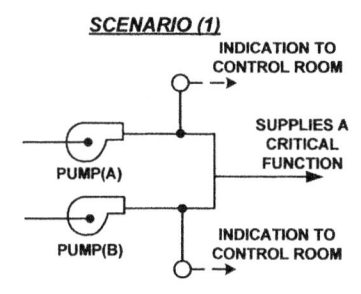

SCENARIO (2)

PUMP(C) — INDICATION TO CONTROL ROOM
          SUPPLIES A CRITICAL FUNCTION

CONDITIONS

1. Pumps (A) and (B) operate simultaneously and either one can supply the required flow to support the critical function.
2. There is an indication of failure for each individual pump.
3. Failure of either pump will not result in an unwanted consequence or an economic concern. Therefore, they are RTF components. However, failure of both pumps will result in an unwanted consequence.

CONDITIONS

1. Pump (C) supplies the required flow to support the critical function.
2. There is an indication of failure for pump (C).
3. Failure of pump (C) will result in an unwanted plant consequence.

SCHEDULED ACTIVITIES:

SCENARIO (1): A CM IS SCHEDULED TO REPLACE FAILED PUMP (A) AND A CM IS SCHEDULED TO FIX AN OIL LEAK ON PUMP (B).

SCENARIO (2): A CM IS SCHEDULED TO REPLACE THE FAILING BEARINGS ON PUMP (C).

QUESTION: WHICH CM TAKES PRIORITY? IS IT ONE OF THE CMs ON THE RTF PUMPS (A) AND (B), OR IS IT THE CM ON THE CRITICAL PUMP (C)?

ANSWER: THE ANSWER IS A VIRTUAL TIE! THE CM PRIORITY IS BASED TOTALLY ON THE UDGMENT AND DECISION PROCESS OF THE OPERATIONS ORGANIZATION IN CONJUNCTION WITH ENGINEERING, DEPENDING ON THE PLANT CONDITIONS AT THAT TIME. WHAT IS THE LESSON TO BE LEARNED? CMs ON RTF COMPONENTS VIE FOR A CORRECTIVE MAINTENANCE PRIORITY EQUIVALENT TO ALL OTHER COMPONENTS, EVEN CRITICAL COMPONENTS, DEPENDING ON THE PLANT CONDITIONS AT THAT TIME.

Figure 3.11 Priority? A CM on a run-to-failure component or a CM on a critical component?

The point I am trying to make is that RTF components are indeed significant and should not be dismissed as being unimportant. Hopefully, this example will shed more light on why an RTF component should not be the orphan of reliability.

3.10 The Anatomy of a Disaster

Nothing cements the importance of what we have learned about the concepts discussed thus far better than showing how the lack of that knowledge can lead to a disaster—in this case, a real-life disaster.

At a major power-generating station, perhaps the single most important system is the steam-driven main turbine and all of its integral components. After all, the turbine is the driver that ultimately generates the electrical output. It is also a rather large unit. In fact, the main turbine is so critical to the operation of the facility that there are a myriad of redundant safety devices to ensure its reliability in case of unforeseen events. In the main turbine design at a specific power plant, there were *three* separate protective overspeed devices to ensure that *triple* redundancy was available to shut the turbine down in the event of an uncontrolled turbine overspeed. A main turbine capable of generating over 1100 megawatts of electrical energy, operating in a runaway, uncontrolled, overspeed situation, should never happen.

The overspeed safety devices were designed such that when an overspeed condition is sensed, the devices activate, sending a signal to the respective steam supply valves to either modulate or reduce steam flow, or completely shut off the steam supply that drives the turbine. Guess what happened? The turbine went into an uncontrolled overspeed. It caused a great deal of damage, and only by luck was a fatality avoided. How did this happen? How could this happen? There was triple redundancy to prevent this occurrence.

Refer to Figure 3.12. It is similar to Figures 3.3 and 3.7, which we analyzed previously in this chapter. Even though there were three separate overspeed devices, they were all in parallel, with no indication of failure for any one individually. Without the proper individual testing of each overspeed device, and without plant personnel understanding of the concept of potentially critical components, the overspeed system did not really have triple redundancy as everyone had thought. *In fact, the protective devices had unknowingly migrated to single-failure vulnerability.* As a result, the turbine went into an uncontrolled overspeed and blew apart, with the turbine blades penetrating the armorlike casing. As can be imagined, the postmortem investigation was not pretty. The hordes of engineering, regulatory, and QA investigators could not conceive of how what they thought to be a triple-redundant system could fail. In fact, from what you have

Fundamental RCM Concepts Explained 63

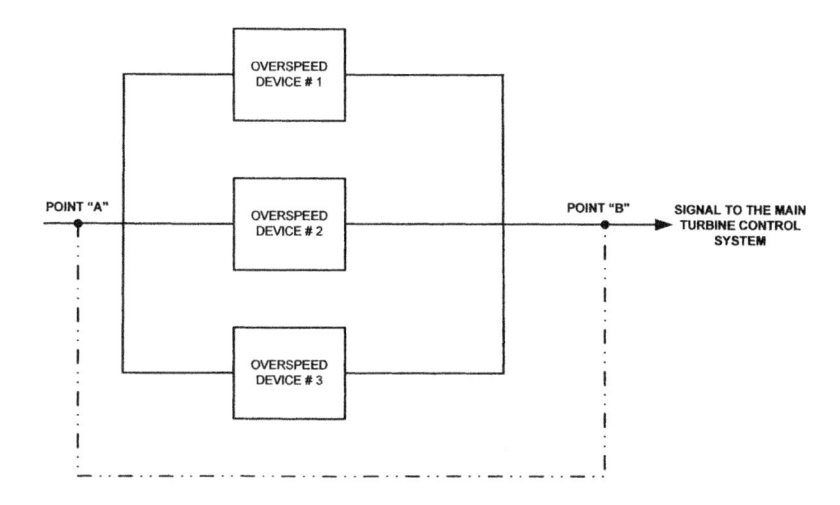

A REAL-LIFE SCENARIO: WHAT IS WRONG HERE?

o The triple redundancy has no individual means of identifying when a failure has occurred to each overspeed device. The individual failure of each overspeed device is not evident. It is hidden.

o The preventive maintenance testing is inadequate since it only verifies the overspeed safety feature from Point "A" to Point "B."

o The "apparent" triple redundancy has been negated to a single-failure vulnerability.

Figure 3.12 An example of triple redundancy, or is it?

learned thus far in this chapter, it should not be surprising to you that this was a disaster just waiting to happen. The "lucky streak" finally ended.

What did the investigation team find? The findings concluded that one of the overspeed devices had been in a failed state for quite some time, as evidenced by corrosion—but no one knew it. The second device evidently had also been in a failed state for some period of time, and this too went undetected. Therefore, that critical overspeed function with triple redundancy was down to single-failure vulnerability and no one recognized it. The third device finally failed. Why didn't the plant personnel know that two of the overspeed devices had previously failed and had been that way for some time? All redundancy was lost, but it went unnoticed. The devices were incorrectly allowed to run to failure. As you now know, in order to be an RTF component the failure must be evident, and obviously there was no individual indication of failure for any of the devices in this case. They were *hidden failures,* which were not understood by the plant staff. The other missed opportunity was how they tested the system. Apparently, it was tested from point A to point B, and no device was tested individually. These were potentially critical components because of their hidden failures, and they were just waiting in their sleeper cell mode to create a disaster. What were the damages? They ran into the hundreds of millions of dollars, and the plant was nonoperational for almost one full year, to say nothing of the lost generating capacity and the commensurate loss of several hundred million dollars in revenue. It was indeed fortunate that there were no fatalities.

How could this have been prevented? Very easily. If you had analyzed each component in the plant and been armed with the RCM concepts you have learned in this book, you would have recognized this vulnerability. It could have been avoided either with a design change to provide failure indication for each device or, more prudently and more cost effectively, with the simple addition of a few PMs to test the overspeed devices individually instead of cumulatively from point A to point B. A very simple task with a miniscule cost of a *few hundred dollars* could have avoided this entire calamity that cost about *half a billion dollars!* That is why I emphasize the importance of the concept of potentially critical components and how identifying hidden failures that are mostly innocuous and not obvious offers the greatest potential for preventing a similar disaster at your facility.

This real-life occurrence illustrates how powerful the concept of being a potentially critical element—in other words, the missing link of RCM—really is. These components were not considered critical because there was no effect with a single failure. In fact, there was triple redundancy, which lulled all of the reliability experts into the misguided logic mode that I mentioned earlier. Therefore, at least two of the three devices were wrongly classified as the equivalent of run-to-failure, and I am not certain that a PM task even existed to cumulatively test these overspeed devices because of the false comfort of believing there was triple redundancy. If they did not even perform a cumulative test, all three devices would have been relegated to an incorrect RTF status. Even if they did test the overspeed devices cumulatively, the third overspeed device failure obviously occurred during an interval in the cumulative test schedule, so it, too, went unnoticed for that period of time.

Because plant personnel lacked an understanding of the missing link and of how to handle hidden failures, the failed overspeed devices totally slipped beneath the radar, and even worse, they were apparently invisible to it. Shortcuts in the RCM process would have had a very slim chance, if any chance at all, to identify this vulnerability.

Now that you have some idea of how important hidden failures are and how important it is to identify them accurately, I believe you are ready for the long version. The following section presents this information about hidden failures, some of which is a reiteration of the preceding discussion. I am doing this by design, so at some point that little light will go off in your head (if it hasn't already) and you will say, "Now I understand the message the author has been trying to convey to me about what RCM and reliability are all about." Don't feel bad if it takes a few readings for this to sink in. Although these concepts are not esoteric, they are sophisticated and subtle. The following information greatly expands on the work of Nowlan and Heap and takes RCM to a whole new level.

3.11 A Deeper Look at Critical Components, Potentially Critical Components, and Hidden Failures—How They All Fit Together

We have briefly discussed the differences between critical and potentially critical components, how to identify them, and how hidden failures affect them. Now let's look a little more deeply into how all this fits together.

I intend to take RCM beyond the traditional methods of establishing a preventive maintenance program. I do not believe you can totally rely on a *cumulative* test for verification of a system or even sections of a system's functionality. Even though the operational test of that entire system may be cumulatively satisfactory, oftentimes undetected failure modes are present that are not immediately evident. These hidden failure modes have the potential to manifest themselves in a plant effect as a result of either time or multiple-failure consequences. These hidden failure modes occur quite frequently and specifically within hidden standby systems or portions of those standby systems.

The reason for identifying any failure mode is to ensure that appropriate PMs are specified if the failure mode(s) could result in a plant consequence, regardless of whether the occurrence of the failure is evident.

1. If the failure mode *is evident,* then a preventive maintenance task can be specified to defend the facility against failure of that component if the failure results in an adverse consequence. These components are identified as *critical* components. It is important to note that the occurrence of the failure mode could become evident either through continuously monitored instrumentation or by the occurrence of the adverse effect itself. This means that if, for example, a valve fails in the closed position and there is associated position indication and an alarm in your control room, that is an indication of failure. When the valve fails closed, an unwanted event can also simultaneously occur, which is likewise an indication of failure—albeit one that we surely want to avoid.

2. If the occurrence of the failure mode *is not evident,* then a preventive maintenance task can still be specified to defend the facility against failure of that component if its hidden failure effect has the potential to result in an adverse consequence. This is why I refer to these components as *potentially* critical components.

Simply stated, a component has a hidden failure mode when either of the following exists:

1. The component's function is *normally active or "in use" when the system is in service,* but there is no indication to control room operating personnel when the function *is lost or ceases to perform.*

2. The component's function is *normally inactive or "not in use" when the system is in service* and there is no indication to the

control room operating personnel that the function *will not be available when needed.*

Whether the component failure mode is identified as critical or potentially critical is subordinate to the main goal of ensuring that, in either case, an applicable and effective PM task is specified to prevent the failure. Remember, too, that an individual component may have several different failure modes that result in different classifications. One failure mode may cause the component to be critical; another may cause the component to be potentially critical or even economic. Another failure mode of the same component may even be classified as run-to-failure. This is typical. However, the final classification of the component *defaults* to the most limiting classification; that is, the component would be classified as *critical*.

To appropriately analyze the hidden failure modes occurring within hidden standby systems, the hidden system must be analyzed in its "demand" mode, since it is not normally operating. Each component in the hidden system must be analyzed as though it has been called upon to function. When you identify a component failure mode that is not evident even upon demand, that is what I call a *true* hidden failure. Many major incidents (such as that of the triple-redundant, normally hidden turbine overspeed system) occur because this circumstance is hardly ever realized and certainly seldom analyzed. Now that you have this knowledge, you can use it to develop your RCM-based preventive maintenance program, and your facility will be safer and more reliable than it was before.

In summary, if a failure *is evident* and if the failure could result in an immediate plant effect, the knowledge of its evident consequence allows you to specify a preventive maintenance task to minimize the exposure to the plant-level effect. Likewise, if a failure *is not evident* and does not result in an immediate plant effect, the knowledge of its potential plant-effecting consequence also allows you to specify a preventive maintenance task to minimize the exposure to a plant-level effect.

Oftentimes it depends on how the function is written as to whether it is a critical or potentially critical component. However, that does not really matter. Remember, you want to ensure that a consequence of failure does not go undetected, and whether or not it is critical or potentially critical, you must still identify a preventive maintenance strategy to prevent failure. For instance, a function can be written in several different ways, so if you wrote the functional failure of the component as "Inlet valve XYZ fails to provide isolation to the main header," it might be a potentially critical failure mode because the failure could be hidden. However, if the functional failure were written as "Given a loss of on-site power, inlet valve XYZ fails to provide isolation to the main header," the failure would be immediate, or critical, since you also included the additional initiating event of the loss of on-site power.

As I mentioned previously, it does not matter how the function is written as long as inlet valve XYZ is identified for a PM strategy to prevent its failure. Whether it is critical or potentially critical, it has been identified for applicability to phase 2, which is the identification of a PM task. The important thing is that the component does *not* escape detection as important and that it has a preventive maintenance task associated with it. As we have also learned, *corrective* maintenance is an integral part of the bigger picture of preventive maintenance.

Now you have an understanding of how the canon law and potentially critical components ensure that no important component gets through the process undetected and that an important component is not inadvertently classified as run-to-failure. Let's see how these concepts work when there appears to be an anomaly.

3.12 Finding the Anomalies

The canon law states that the failure must be a single failure; it must be evident; and it must be fixed in a timely manner to qualify for an RTF classification. However, if the component is so far down on the hierarchy of relative importance, although the component failure might be *evident,* there still may not be any plant consequence even if there were a myriad of additional failures in conjunction with the original failure. When the logic delineates a multiple-failure scenario that nevertheless does not have any unwanted consequence of failure, you should always ask, "Why is the component even installed in the plant?" Another possible anomaly exists if those components are installed in your facility strictly for convenience or if they obviously have very insignificant value. It is not uncommon to find that even though such components have failures that are *not evident,* there nevertheless may not be any plant consequence even if there were a myriad of additional failures in conjunction with the original failure.

For example, in a manufacturing facility, the water supply to the lavatories may indeed have hidden multiple failures and still be classified as RTF. However, what about the service water system supplying lavatory service water on a B747 for a 12-hour flight to Australia with over 300 people on board? The lavatory water supply would most definitely not be classified as RTF on a B747.

The inherent benefit of the concepts of potentially critical components and the canon law is that they make these anomaly components *stand out and be noticed* so that nothing of importance escapes the RCM logic. These concepts provide a path for exception, but only after that exception has been carefully analyzed. Any anomaly components can then be evaluated for whether they should continue to be maintained or whether they should be considered for removal from the plant entirely; however, that decision is yours.

An amusing experience exemplifies the need for understanding all this. A senior management person once asked me why a certain room heater was classified as a critical component and not an RTF component since nothing happened when it failed and it didn't matter when it was fixed. The individual asked me this question in the month of August when it happened to be rather warm and a room heater was definitely not needed. I explained that while his logic prevailed in the summer months, it did not prevail in the winter months, when there was a regulatory requirement to maintain the temperature of the room, which contained critical equipment, above a minimum of 65°F. As the canon law also states, "Corrective maintenance is commensurate with all other components depending on the plant conditions at that given time." So in the summertime, corrective maintenance of the heater might not be a top priority, but in the winter months it would be. As simplistic as this may seem, this is oftentimes the level of understanding of RCM even among senior management.

3.13 Failures Found During Operator Rounds

Another factor in the equation regarding whether or not the occurrence of the failure is evident pertains to the "formal" operator rounds performed at your facility.

There are many instances where the component failure status is not evident to your control room personnel via continuously monitored instrumentation but will become evident during an operator round. Operator rounds are an integral part of a preventive maintenance program and are considered a preventive maintenance "inspection" type of task.

As long as the operator round is contained in a formalized procedural process and takes place at discrete time intervals, credit may be taken for calling that failure *evident*. Every facility is not like a commercial jet aircraft, where virtually everything is monitored in the cockpit, or a nuclear power plant, where virtually everything is monitored in the control room—except of course, hidden failures such as the one that caused the turbine disaster.

3.14 Redundant, Standby, and Backup Functions

There is much confusion about *redundant, backup,* and *standby* equipment. Is failure of the redundancy evident? In many instances, it will *not* be evident. For example, in most instances redundancy exists in the form of a backup or standby function. The differences between redundant, standby, and backup functions are widely misunderstood and are a source of confusion even within the nuclear industry. Figure 3.13 represents the most typical examples of these functions. Also refer to Figure 3.14, Scenario 4 (scenarios 1 through 3 will be discussed later).

A major source of misunderstanding arises when there are two pumps, for example, and the function of one of the pumps is that it be used in a standby or backup mode for the other. Thus, only one is operating at a given time, and the other is functioning as a standby or backup should the first one fail; either pump can act as the backup for the other.

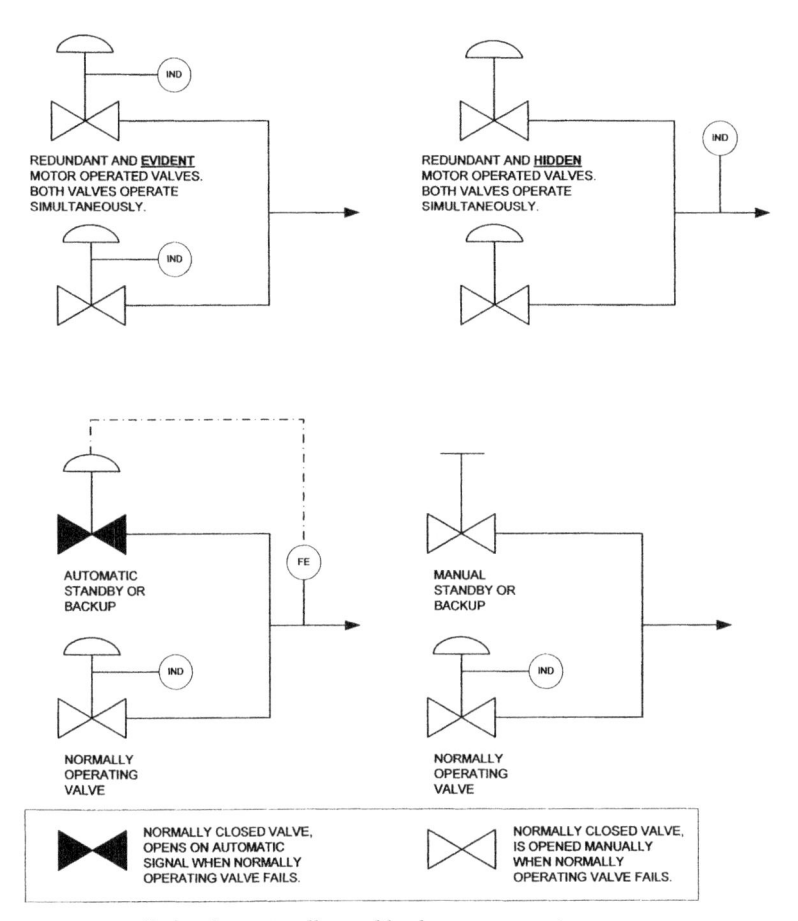

Figure 3.13 Redundant, standby, and backup components.

When a component is called upon to perform a *standby* or a *backup* function, it must be assumed that the normally operating component has failed. Thus, if the backup should fail, an unwanted consequence could occur, and that is definitely not the time to find out that the standby or backup component does not operate. Therefore, *standby* or *backup* components are considered critical, and a PM is required (assuming, of course, that an unwanted consequence of failure will occur). Since either pump may act as the standby or backup for the other, they would both be critical.

72 Chapter Three

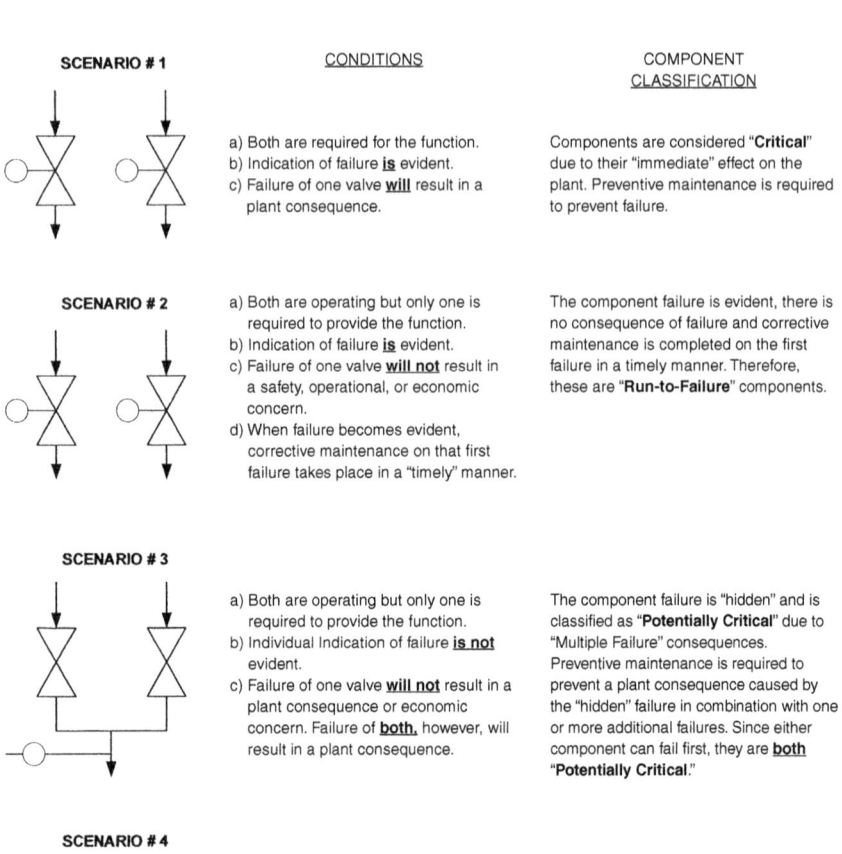

SCENARIO # 1	CONDITIONS	COMPONENT CLASSIFICATION
	a) Both are required for the function. b) Indication of failure **is** evident. c) Failure of one valve **will** result in a plant consequence.	Components are considered "**Critical**" due to their "immediate" effect on the plant. Preventive maintenance is required to prevent failure.
SCENARIO # 2	a) Both are operating but only one is required to provide the function. b) Indication of failure **is** evident. c) Failure of one valve **will not** result in a safety, operational, or economic concern. d) When failure becomes evident, corrective maintenance on that first failure takes place in a "timely" manner.	The component failure is evident, there is no consequence of failure and corrective maintenance is completed on the first failure in a timely manner. Therefore, these are "**Run-to-Failure**" components.
SCENARIO # 3	a) Both are operating but only one is required to provide the function. b) Individual Indication of failure **is not** evident. c) Failure of one valve **will not** result in a plant consequence or economic concern. Failure of **both,** however, will result in a plant consequence.	The component failure is "hidden" and is classified as "**Potentially Critical**" due to "Multiple Failure" consequences. Preventive maintenance is required to prevent a plant consequence caused by the "hidden" failure in combination with one or more additional failures. Since either component can fail first, they are **both** "Potentially Critical."
SCENARIO # 4 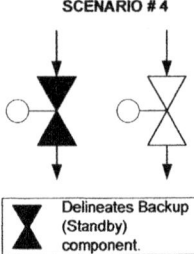 ■ Delineates Backup (Standby) component.	a) The "backup or standby" component is required to provide the function in the event of failure of the operating component. b) Either component can function as the backup (standby) for the other component. c) Indication of failure is evident for both components.	When a component is called upon to operate in a "Backup or Standby" function it is assumed that the normally operating component has failed. If the Backup (Standby) should fail, a plant consequence would occur. Therefore, the Backup (Standby) component is considered "**Critical**" and preventive maintenance is required to prevent its failure. If either component can act as the Backup (Standby) for the other, they are **both** "Critical."

Figure 3.14 Typical examples of component classifications.

Redundant components or redundant systems usually operate simultaneously. If there is individual indication that each one is operating, and likewise there is individual indication when each one should fail, the *redundancy* allows for a run-to-failure classification in the absence of any regulatory, economic, self-defined, or other operational considerations.

Automatic backup or automatic standby components do not operate simultaneously and will normally start on an automatic input signal—for example, a signal caused by a high or low pressure, a high or low temperature, a loss of flow, or a high or low liquid level signal. *Manual backup* or *manual standby* components also do not operate simultaneously and will start or become operationally functional only by means of a manual input rather than an automatic one. In either case, if an unwanted plant consequence should occur as a result of failure of the backup or standby component, that component would be considered critical. The majority of people, including many experts, still believe that just because there *appears to be redundancy,* this automatically qualifies the components as RTF. Totally wrong!

3.15 Typical Examples of Component Classifications

Figure 3.14 coalesces the typical examples of critical, potentially critical, and run-to-failure components so that the subtleties become apparent. As noted in Figure 3.14, there is a vast difference between a component operating in a backup function and one that is not. In scenario 2, of Figure 3.14, the component is a *run-to-failure* component, while the component in scenario 4 is *critical.*

3.16 Component Classification Hierarchy

The classification hierarchy is shown in Figure 3.15. As mentioned previously, different component failure modes for the same component may have different classifications depending on the consequence of their failure. Differences will *default* to the most limiting classification for that component. Remember, too, that a single component will oftentimes have numerous different types of PM activities associated with it, but each PM activity maintains its own classification of importance as to whether it is critical, potentially critical, commitment, or economic. There are no PMs for RTF components.

> **Important Notes:**
>
> - A component will have several different failure modes resulting in different classifications for each failure mode. The *final* component classification *defaults* to the highest classification.
> - Unlike components, the classification of a PM task *remains* at the classification level determined by the specific failure mode.
> - Refer to *Section 3.16* for a comprehensive explanation and the reasons for the above logic.
>
> | Component Classification Hierarchy | PM Task Classification Hierarchy |
> |:---:|:---:|
> | Critical | Critical |
> | Potentially critical | Potentially critical |
> | Commitment | Commitment |
> | Economic | Economic |
> | Run-to-failure | Run-to-failure |

Figure 3.15 Component classification and PM task classification hierarchy.

Unlike the component classification, the PM classification *remains* at the same level; it does not default to the highest level. This means that if a *component* is classified as critical but its associated PM requires it to be painted every year, that PM to paint it could be an economic one. Again, I include this explanation to ensure total accuracy, thoroughness, and correctness of the process. The instances where the PM classification for a component varies from either critical or potentially critical to economic are not that frequent. However, you should be aware that this could occur. Therefore, there could be a critical component with an economic PM task. I differentiate this because even though it doesn't happen that often, can you imagine what a craftperson would think if he or she was assigned a PM task to paint the exterior of a component and that PM was classified as critical to the plant? It would completely detract from the relative importance of performing work, which is precisely what you are trying to achieve via RCM. I am presenting RCM in a manner that makes sense and have gone to great lengths to keep it simple but at the same time sensible and correct.

3.17 The Defensive Strategies of a PM Program

I like to compare a preventive maintenance program with the defensive strategy of a football team. When you think about it, they are quite similar. The defensive objective in football is to prevent the opponent from doing something negative or damaging to your team. The objective of a PM program is also to prevent the opponent, in this case the equipment, from doing something negative or damaging to your facility. In football, the defensive line acts as the first line of defense, then the linebackers, and then the defensive safeties. In my approach to RCM, the critical components are the first line of defense, then the potentially critical components, then the commitment and economic components.

The *first line of defense* for protecting your plant against unplanned equipment failures consists of identifying *critical* components. When these components are called upon to function, a single failure immediately results in a detrimental plant consequence.

The *second line of defense* for protecting your plant is to identify *potentially critical* components. When these components are called upon to function, their failure is hidden and does not result in an immediate plant consequence. However, either over time or in combination with one or more additional failures or initiating events, a detrimental plant consequence will occur.

The *third line of defense* to protect your plant is to identify *commitment* and *economic* components. When these components are called upon to function, their failure is not critical or potentially critical, but they result in either a missed commitment or an economic concern.

3.18 Eliminating the Requirement for Identifying Boundaries and Interfaces

In Chapter 2 and at the beginning of this chapter, I made a startling statement: that RCM does not require the identification of system boundaries and interfaces. This is an extension of the philosophy of the pioneers of RCM, Nowlan and Heap. Not requiring the identification of system boundaries and interfaces is a primary aspect of *RCM Implementation Made Simple*. RCM was intended to be a structured decision process to identify those items whose failure is significant at the *component level*. It is the functions and failure possibilities at the component level that are important for preventing plant-level consequences.

There are several ways to commence an RCM analysis. The only method currently espoused in any publication about classical RCM is to identify functions at the system level and then at the subsystem level. That is not an incorrect method; however, it is a very unwieldy and cumbersome method because you are forced to identify system boundaries and interfaces. That is perhaps the most time-consuming and complicated aspect of the entire RCM process. It also leaves you vulnerable to missing some very important component functions.

To identify system boundaries, you must compartmentalize every system and every subsystem and then identify the interface *beginning* and *end* points, where the adjoining systems and subsystems reside. This must be clearly marked and identified on the plant schematics. You must know where one system ends and the other begins or you may have countless components fall through the cracks and not get analyzed. This does not mean just identifying the boundaries and interfaces at the system level; it also involves identifying the boundaries and interfaces at the subsystem level within that overall system.

There are literally hundreds of arbitrary interface points that must be identified for each system. This means that thousands of individual arbitrary components must be specifically identified and documented as interface points. Then there are the interfaces of power supplies, fluid transfer, signal inputs, and so on. This is quite a daunting task, even for experienced RCM analysts. But that is still not the end of the process of identifying boundaries and interfaces. Often a valve becomes one of the interface points, in which case you need to decide whether that valve controls the flow from the adjacent system into the system being analyzed. If it does, it is documented as an out-system in-interface and the valve belongs to the adjacent system. If the valve controls the flow from the system being analyzed into the adjacent system, it is documented as an in-system out-interface and it would belong to the system being analyzed. Also, keep in mind that a specific component can reside in only one subsystem of the larger system. By the time you have identified all of the boundaries and interfaces, you could be well on your way toward completing the entire RCM analysis at a midsize facility.

As Chapter 2 mentioned, identifying boundaries and interfaces is one of the primary reasons for the failure of an RCM program to be implemented. Perhaps it might be preferable to go through these machinations for a nuclear plant or commercial aircraft, but they are still not necessary because they are all captured anyway when identifying functions at the component level. *All of the interface points will ultimately be included as part of the process anyway when you identify functions at the component level, so you do not have to specifically research thousands of component I.D.s just to identify beginning and end points.* Every component must be analyzed no matter how big or small it may be. Additionally, every function of each component must be analyzed. This is a straightforward process and is relatively easy; it does not have to be accomplished in any sequence as you would need to do if you started by identifying functions at the system level. Eventually you will have completed the analysis for all components, and, after all, it is the component failure effect that we are seeking.

The majority of facilities will almost certainly have a relatively smaller population of equipment to analyze than a nuclear plant or a commercial jet aircraft. A single average-size nuclear plant system will probably have more components to analyze than an entire average-size facility in another industry. Therefore, it is much easier to define functions directly at the equipment level and eliminate all of the unnecessary burdens of identifying boundaries and interfaces. The time and expense saved will be substantial.

3.19 Functions and Functional Failures Are Identified at the Component Level, Not the System and Subsystem Level

The whole reason for identifying functions at the system level was to be able to collectively identify numerous individual components that could cause the functional failure of the system. There is a benefit to doing it that way, albeit a rather flimsy one.

Even when identifying functions at the *system level,* the intention is still to get down to the *component level* to determine those *component failure modes* that have an adverse effect on the safety and reliability of your plant. If all of the effort is to get us to the component level anyway, why not start there to begin with? And that is precisely what *RCM Implementation Made Simple* does.

It has been my experience that the difficulty and time involved in identifying boundaries and interfaces are not worth the effort and do not buy you any more accuracy. In fact, unless you are a real RCM expert and have previously implemented a large-scale RCM program, identifying functions at the system level may even result in *less accuracy* in the analysis. What do I mean by that? Even for some of the most experienced analysts, it is very difficult to identify every system and subsystem function. Frequently, stray components are left out and not accounted for by any of the identified system and subsystem functions. That could be a dangerous omission.

Another negative consideration in this process is that many of the different functions of a system/subsystem identify the same equipment as having an impact. This repetition is eliminated by identifying functions at the equipment level. The functions for each component are analyzed only once, thus there is no repetition.

Even though identifying functions at the equipment level obviously involves identifying more functions than if they were identified at the system level, the time saved and the level of accuracy achieved are well worth it. No component is left out of the analysis—there are no stray components. Performing the analysis is also much more straightforward and easy to understand. The analysis results are likewise easier to implement, and any subsequent additions to the program are simpler to make. I will show you how to use this simple approach to RCM in Chapter 5. Some may prefer to identify functions at the system and subsystem level, thereby identifying boundaries and interfaces; I will also explain in Chapter 5 how to develop an RCM program using this more difficult approach.

Every system, no matter how complex, can be broken down into its simplest elements, which exist at the component or the equipment level. The terms *component* and *equipment,* as used in this book, are synonymous. The component level is that level where a separate equipment identification number (equipment I.D.) is specified. It is the level for determining the function of a valve, motor, pump, switch, heat exchanger, circuit breaker, and so on. This level does not include the subassembly piece parts such as the bearings, armature, stator, shaft, and crank arm. Subassemblies become important when identifying the causes of component failure, and they are included as part of the process when we get to the PM Worksheet in Chapter 6. So for determining functions, we remain at the equipment level.

Another very important consideration when identifying functions at the equipment level is new or modified equipment or different operating schemes whereby changes to the analysis would be required. Making these changes at the equipment level is so much easier than having to go into the functions at the system and then the subsystem level to introduce them.

Identifying functions at the equipment level also serves as an excellent training program for those involved in the analysis to acquire detailed knowledge about how the plant operates. You might be surprised at how much you can learn from this experience that you did not know before. Remember, too, that RCM is a living program, and it is much easier to make living program changes and adjustments at the equipment level.

Identifying functions directly at the equipment level could be a revolutionary element in classical RCM. The time savings are enormous, the analysis is much simpler, changes and additions are easier to make, the process itself is much easier to understand, and above all else, the analysis is more thorough and accurate.

3.20 The Quest for the Consequence of Failure

The *consequence of failure* is what the whole RCM process is all about. Everything you do in RCM and every part of the analysis are driven to obtain only one answer: *what is the consequence of failure?* Once this is determined, you need to figure out how to prevent the unwanted failure consequence via a prescriptive preventive maintenance program that includes a vast selection of PM tasks from which to choose.

How do you get there? The chronological sequence is as follows:

1. Start by identifying all of the functions of the equipment.
2. Identify the functional failures or the different ways the function can fail to be provided. (While I do not necessarily subscribe to this step in the process, it is not that difficult to identify the functional failures, since they are the exact opposite of the function. I have included this step primarily to satisfy SAE Standard JA1011. This is explained in Chapter 4.)
3. Identify the different failure modes or the different ways the equipment can fail, which will result in the failure of the functions identified previously.
4. Identify the effects of the failure modes. This could be at the local level, the subsystem level, the system level, or the plant level. From experience, I have found that any real value in identifying the effect of failure is at the plant level. However, as an intermediate informational acknowledgment, the effects at the system level are also identified.
5. Identify what the consequences to your plant or facility will be as a result of the failure effects. These consequences are based on what you deem important and appropriate based on the asset reliability criteria that you establish. (This is discussed in Chapter 5.)

All of these steps are part of phase 1, as delineated in Section 3.1.

Phase 1: Consists of *identifying* equipment that is important to plant safety, generation (or production), and asset protection.

This brings us to Phase 2, which is the specification of the different types of PM tasks.

Phase 2: Consists of *specifying* the requisite PM tasks for the equipment identified in phase 1. These tasks must be both applicable and effective.

This phase also includes a very prescriptive method for identifying default actions if an applicable and effective preventive task cannot be found. Chapter 6 looks at developing the PM task strategies in detail.

All of the preceding steps align with the seven steps specified in SAE Standard JA1011 for determining what constitutes an RCM program. This is further delineated in Chapter 5.

The third phase is the execution of the tasks.

Phase 3: Consists of properly *executing* the tasks specified in phase 2.

It is important to note that the question "What is the consequence of failure?" is *totally independent* of the pedigree of the component. By this, I mean the consequence of failure is independent of whether or not the component is safety-related or non-safety-related, or whether it has a lower-quality class rating than other components, or whether it is located in an area of apparently less importance in your facility, or whether it seldom results in corrective maintenance compared to other components, or whether it has a lower probability of failure (if you are statistically inclined), and so on. The question for each failure mode of each component is "What is the consequence to the facility if it should fail?"

As for probabilities of failure, those statistics may be used for determining the periodicity of a specified task, but they are not a factor in determining whether to include a PM task as long as the failure mode is a credible one. Credible failure modes are explained in Chapters 4 and 5, and the aspect of the analysis process for specifying PM task periodicities is discussed in Chapter 6.

3.21 The COFA Versus the FMEA

The commonly used term for the analysis format of the RCM process is *FMEA*, which stands for *failure modes and effects analysis*. Some programs even call it *FMECA,* or *failure modes, effects, and criticality analysis*. Use of these analysis formats is not incorrect; they are both valid. However, I have found that the identical information can be collected, but with much more conceptual clarity, by employing the *COFA*, which stands for the *consequence of failure analysis*. I developed the COFA to include all of the same attributes as the FMEA as well as additional attributes, so that it is all-encompassing—and it does this at the component level. It is the component level that is the final destination for arriving at the consequence of failure. After all, both the FMEA and the FMECA drive you to identify the consequence of failure for each component failure mode, so why not be more specific, accurate, and clear by using a more precise format and calling it what it really is?

I have *integrated* all of the RCM logic for the identification of critical, potentially critical, commitment, and economic components within the COFA framework. Furthermore, I have included the deterministic logic for whether the occurrence of the failure is evident or not. The COFA also includes the decision process for determining the consequence of failure based on the asset reliability criteria specified. The COFA thus constitutes a simplified and self-contained, all-inclusive RCM logic analysis, which is much more straightforward and comprehensive than the FMEA. This is truly *RCM Implementation Made Simple*. Chapter 5 guides you through each step of the COFA Logic Tree and the COFA Worksheet, with examples of an actual analysis.

Another feature of the COFA is that it maintains the distinct separation of the process for *defining* critical, potentially critical, commitment, and economic components from the process of *specifying* their associated applicable and effective PM tasks. Why? For clarity and simplicity. I have found that trying to figure out what types of PMs to specify at the same time you are trying to define the criticality of a component only adds an element of confusion to the process. It also detracts from the mindset and momentum of phase 1, which is to define the equipment first, before you embark on specifying the tasks for that population of equipment.

There is another valid reason for doing it this way. As I mentioned in Section 3.2, you must identify the functions of each component *individually*. You *cannot group* similar equipment types together to identify their functions; however, you *can group* similar equipment types together to specify the different kinds of PM tasks. In fact, this is the preferred way to complete that part of the process. Why? For example, there may be (and probably will be) many similar types of motors, pumps, switches, motor-operated valves, and air-operated valves, which may employ the same type of preventive or predictive maintenance applications. Therefore, it is much more prudent and efficient to specify the PM tasks by equipment type.

You may even wish to group similar equipment types together that were classified as economic components. This might afford you the opportunity to specify a different, less stringent, set of preventive and predictive applications than you would specify for the critical and potentially critical components of that type. This is another reason that I distinctly specified the identification of the different component classifications; in addition to the benefit of prioritizing work and optimizing available resources, this can also be used to establish the degree to which you decide to specify prescriptive PM tasks for your strictly economic components.

Remember, the *periodicities* of the preventive and predictive tasks are still determined on an individual component basis, using several decision factors—the component's environment, failure history, process flow, temperatures, and so on. Chapter 6 discusses the various factors for determining the correct periodicity for a PM task.

3.22 How Do You Know When Your Plant Is Reliable?

How do you know when your plant is reliable? If you have not performed an RCM analysis, and you are satisfied with your present reliability achievements, perhaps you have been running on luck. Or perhaps your plant is relatively new and has not yet been subject to wear-out concerns. The first of these two possibilities is not acceptable. The second offers only short-term comfort, which allows you to exist in an illusory state of complacency that is only temporary at best.

I should also warn you about another common situation: a state of illusion where *unplanned* failure events occur and each event is neatly packaged as an "isolated" incident. After several dozen "isolated" incidents, one begins to believe that something more sinister is occurring in regard to the reliability of your facility. If you participate in some type of periodic inspection program required by a regulatory body such as the National Aeronautics and Space Administration (NASA), Federal Drug Administration (FDA), OSHA, EPA, NRC, or FAA, you should be aware that these agencies are becoming more alert to the packaging of "isolated" incidents as a camouflage to conceal unacceptable reliability performance.

Quite often (and unfortunately so), reliability becomes the "flavor of the day." When some major unwanted event happens, there is a rush to bolster the reliability of that specific piece of equipment with a myriad of preventive maintenance tasks, new designs, and everything in between. This obviously appeases senior management, allows for the inevitable follow-on root-cause failure report to document corrective actions, and probably does some good for the equipment in question, but crisis management is no way to run an airline, a nuclear plant, a missile program, a cruise ship, a manufacturing plant, an oil refinery, a paper mill, or *any* type of facility.

As we will learn in later chapters, there are no guarantees that a preventive maintenance program alone—even a premier RCM program—will prevent each and every failure, any more than all of the safety features built into an automobile will guarantee against sustaining an injury in the event of a collision. But there are guarantees that some automobiles offer a greater degree of protection against injury than others, and this logic also holds true for a premier RCM program. Such a program provides for a far greater degree of reliability and protection against catastrophic occurrences than if it were nonexistent.

Once you have performed a comprehensive RCM analysis and there is a very clear absence of any serious consequences of failure at your plant, and your workload is primarily utilized for *planned* maintenance activities rather than *unplanned* events, that is somewhat of an indicator of a healthy reliability program.

Chapter 9 shows you how to establish a monitoring and trending performance indicator to quantitatively measure your reliability performance.

3.23 Chapter Summary

We have covered a lot of ground in this chapter. Even if you have some RCM expertise, certain concepts discussed here will be totally new to you. If you are an RCM novice, all of these concepts will be new to you. So let's summarize what we have just learned about these important RCM concepts and principles.

- There are three phases to an RCM program. The first phase is to define the equipment population that will be part of your preventive maintenance program. The next phase is to specify the PM tasks for the identified population. The third phase is to execute the specified tasks.
- RCM is a single-failure analysis except when the single failure is hidden. It then becomes a multiple-failure analysis.
- When a component is required to perform its function and the occurrence of the failure is *not* evident to the operating personnel—that is, the immediate overall operation of the system remains unaffected in either the *normal* or the *demand* mode of operation—then the failure is defined as being *hidden*.
- A multiple-failure analysis is *required* when the occurrence of a single failure is hidden. Addressing hidden failure modes is one of the key aspects for achieving plant reliability.
- Critical components are immediately evident when they fail, and the unwanted consequence of this failure is immediate.
- Potentially critical components are not immediately evident when they fail, and they entail no immediate consequence of failure. They are hidden failures and become critical either by *multiple failures,* by the duration of *time,* or by other *initiating* events.
- Potentially critical components are the *missing link* in RCM. They are a key to reliability success.
- Analyzing potentially critical components probably offers the greatest degree of protection against a major disaster at your plant.
- Preventive maintenance testing should be performed at the component level and not solely at the system or subsystem level.

- The overwhelming majority of scenarios involving the discovery of potential plant vulnerabilities can be very easily resolved with a supplemental PM activity added to the preventive maintenance program. The need for a design change is the exception rather than the rule.
- Components associated with standby safety systems such as emergency backup power, fire protection, emergency cooling, and other such systems that are not normally used must be analyzed in their *operating or demand* mode; that is, the analysis can only be performed assuming the applicable standby safety system has been called upon to function.
- Commitment components are associated with commitment requirements and are also *usually* classified as either critical or potentially critical.
- Economic components have no *safety* or *operational* consequences as a result of failure. Their failure results only in the cost of labor and/or materials for their restoration or replacement.
- Run-to-failure components or systems *must* follow the *canon law* as defined in Section 3.8 and the RTF philosophy as shown in Figure 3.9b. Run-to-failure is a vastly misunderstood concept.
- *Corrective* maintenance is an integral part of the bigger picture of a preventive maintenance program.
- Anomalies involving any of the RCM concepts offer the opportunity for the component to stand out and be noticed for further review and an evaluation of whether the component is even needed in the plant.
- There are differences between redundant, standby, and backup components.
- The typical examples of component classifications in Figure 3.14 show how subtle differences in the design of a system and whether or not failure indication is individually evident can result in very different classifications.
- The defensive strategies of a preventive maintenance program are similar to the defensive strategies of a football team. The aim is to protect your asset against an unwanted consequence.

Fundamental RCM Concepts Explained 87

- Establishing and identifying boundaries and interfaces are *not* a required part of the classical RCM process.
- The component (or equipment) level is where we analyze for functions. The piece parts of the equipment are part of the process when we analyze for the causes of failure (which are discussed in Chapter 6). The component level is where a separate equipment identification number (equipment I.D.) is specified.
- The COFA is a much more simplified, accurate, and comprehensive method for identifying equipment functions and their consequence of failure than the FMEA or FMECA format.
- *Functions* for similar components *cannot* be grouped together. *PM tasks* for similar components *can* be grouped together.

Some additional governing rules for an RCM analysis include the following:

1. Any component failure that can result in a personnel safety issue is always deemed critical.
2. The RCM analysis is applicable to active (nonpassive) components *unless* equipment failure histories or other justifications indicate that they should be included. Passive (nonactive) components include such things as structural components or metal plates, flow orifices, and floor grating. Piping wall thickness wear is covered by programs such as flow accelerated corrosion (FAC) analyses. Maintenance of structural members is covered by separate structural inspection programs.
3. Manual valves are not normally included in the analysis *unless* there is a history of the valve needing periodic preventive maintenance or if the manual valve functions as a backup or standby component for a critical or potentially critical component.
4. RCM applicability for instruments is discussed in Chapter 7.

Now let's look at some of the tools you need to commence the analysis.

Chapter 4

RCM Implementation: Preparation and Tools

In this chapter, you will learn about the tools and the preparation needed to commence your analysis. The tools include simplified spreadsheets that I have designed to expressly maintain the simple flow of logic for the analysis. The analysis itself and all of the detailed explanations for the analysis logic are discussed in Chapters 5 and 6.

Not unlike carpenters who need to have all of their tools before they begin construction on a project, you will need some RCM tools, too. Also like carpenters, you can use very simple tools or very complex and expensive ones. This chapter introduces you to all of them, but the choice is yours as to what tools you decide to employ. Remember, the simple ones usually get the job done just as well as the expensive ones. However, you will find that the larger and more complex your facility, and the more individual equipment I.D.s you have, the more sophistication you will require from your tools.

For example, an Excel spreadsheet and a 3-by-5 index card filing system may not be appropriate for a nuclear plant, but they might be quite suitable for a small to midsize manufacturing facility. The sophistication of the tools used for storing and retrieving data also depends solely on your needs and the complexity and size of your facility. If your plant is governed by a regulatory agency, you may have to justify your maintenance program actions to the applicable regulators, showing how you arrived at your decisions; thus, you will probably need a more elaborate computerized maintenance management system (CMMS) and support from your information technology (IT) organization. The internal requirements for the management of data at your plant will be determined by your own knowledge and experience, which will ultimately establish the level of sophistication you will need. The various options are discussed later in this chapter.

4.1 Preparation

To prepare for the implementation process detailed in Chapters 5 and 6, I recommend that you assign an RCM single point of contact (SPOC) or an RCM champion or coordinator to gather your RCM representatives together from maintenance, operations, and engineering. Craft representatives should also be members of the team. These representatives can rotate from within the different departments, depending on the specific knowledge requirements at the time. All members will need to know what your goal is and how you plan to achieve it. The implementation process described in Chapters 5 and 6 should be well understood by the department representatives. Your RCM coordinator does not need any special facilitator training or RCM expertise. He or she should, however, be capable of understanding *RCM Implementation Made Simple* and be an enthusiastic leader of others. The RCM coordinator should also understand the COFA logic and be able to explain it to the other team members.

I created the COFA as a pivotal fundamental aspect of *RCM Implementation Made Simple*. I did this so that dozens of components can be analyzed in a very short time. If the team representatives were to meet once per day for two to three hours each day, or even better, if the team could go off-site for one week at a time to analyze hundreds and perhaps thousands of components each week, you would find that you could complete the COFA in a relatively short time, ranging from a few weeks to a few months, obviously depending on the size of your facility and the number of components you have. Most RCM programs are measured by the number of years it takes to complete the program.

A time element for completing an RCM analysis that is measured in years is not only inappropriate, it is unacceptable.

Once the member representatives have been selected and the *mission statement* has been thoroughly discussed, there are some fundamental elements that are an absolute requirement.

4.2 The Sequential Elements Needed for the Analysis

Whether you have a small facility, a midsized facility, or a major complex facility, the following represents a logical approach for commencing your analysis. It consists of the following sequential elements:

1. A simple but comprehensive alphanumeric equipment database must be established.
2. All pertinent informational resources must be available.
3. A descriptive convention for specifying functions and functional failures must be specified.
4. An Excel COFA spreadsheet (or more sophisticated RCM software if you prefer) must be available. An example of the COFA Excel Worksheet is shown in Figure 4.2 and is discussed later in this chapter.
5. An Excel PM task spreadsheet (or more sophisticated RCM software if you prefer) must be available. An example of the PM Task Worksheet is shown in Figure 4.3 and is discussed later in this chapter.
6. An Excel Economic Evaluation Worksheet must be available. An example is shown in Figure 4.4 and is discussed later in this chapter.

Let's look at each one of these in detail.

4.2.1 A simple but comprehensive alphanumeric equipment I.D. database

This should already be available, and if it isn't, it needs to be. You have to start with this database. All the equipment in your plant should be included in your plant design drawings, piping and instrumentation drawings (P&IDs), facility schematics, electrical drawings, system operating manuals, training manuals, and so on. As we learned in Chapters 2 and 3, system boundaries and interfaces are not necessary for the analysis; nevertheless, your equipment is usually contained in some kind of a system format. If you have the capability and if you prefer to sort your equipment I.D.s by a system tag number, that sort capability will be helpful if you should want to view your performance on a system basis. In Chapter 9, we develop a monitoring and trending program to accomplish that; however, keep in mind that it is the component level where any questions arise relative to the analysis.

Therefore, some type of labeling by system is useful to have, but if your facility is not designed that way and it is too much of an effort to create a simple system compilation of equipment I.D.s, don't spend the time and effort to create one. The analysis does not require it because system boundaries and interfaces are not needed. Only a comprehensive alphanumeric equipment I.D. database is needed.

My experience has shown that whenever a question arises in regard to the RCM analysis, it is *always* at the component level. The question universally asked is "What is the classification of component XYZ?" Or "Why was component XYZ classified as *critical, potentially critical, commitment, economic, or RTF?*" There are no classifications at the system level, and any data retrieval requirements of the analysis relative to failure modes and consequences of failure are always by equipment I.D. That is one of the primary reasons I developed the COFA.

An interesting inquiry that often arises concerns an electronic black box, such as a controller or a power supply, that has a myriad of internal electronic parts including microchips, diodes, resistors, and all kinds of circuit boards. Although there are very good testing procedures for an electronic black box as a component itself, there is still the possibility that an electronic piece part will fail without any prewarning. However, it is not practical to perform an individual RCM analysis on the thousands of internal electronic piece parts of a black box. Nonetheless, if you have a failure history for a specific piece part, such as a specific electrolytic capacitor within a black box, you should include replacement of that capacitor via a time-related PM. The same logic applies to other piece parts or passive components. If they have a history of their failure consequences, this is always justification for including them in your maintenance program.

At this point in the compilation of your equipment database, remember that it is important that no equipment I.D. be overlooked, regardless of what you may initially think about its criticality. The COFA analysis will determine that. It is quite common that the COFA will reveal a component to be significantly more important than you assumed it to be, and that component could easily have been overlooked. Another reminder is that RCM is not a pick-and-choose option. Each component and each function of the component must be analyzed. Beginning your analysis by dissecting your existing PM program and then trying to justify what you are already doing is *not* classical RCM. In fact, it is not any form of RCM.

The equipment database should include components only at the equipment level, where a specific I.D. number has been assigned, such as valve, motor, pump, or compressor. Piece parts are not a part of this alphanumeric equipment I.D. database. Piece parts are, of course, important and need to be included in a master equipment list, but they are not at the level where the equipment functions are identified and therefore are not a part of the equipment I.D. database. Piece parts are discussed in Chapter 6.

4.2.2 Informational resources

Depending on the level of research you require, you may wish to pull together the following resources: applicable maintenance, operations, and engineering procedures, training manuals, study guides, design basis documents, vendor manuals, technical bulletins, plant equipment drawings and schematics, electrical schematics, P&IDs, and so on. The informational resources should contain the requisite information for ascertaining the functions of the equipment and the ability to determine what the consequences of failure would be at the system and plant level should the equipment fail to function. In addition to your various design drawings, plant schematics, and other resources, the expertise and knowledge of your maintenance, operations, and engineering personnel will provide an additional valuable input to this element.

Many industries share equipment reliability data with each other. If your industry shares failure data about similar types of equipment, that is a very good source for obtaining information. This informational resource would include failure modes and failure causes for similar equipment that you have in your facility but which may not yet have experienced those failure modes.

4.2.3 Establishing convention

The COFA requires that you define the different failure modes for the functional failures. To simplify the phraseology of the failure modes I recommend that you establish your own convention. By that I mean if a valve fails to open, it could also be written as "valve fails closed." It doesn't matter what convention you use, but you should select one and stick with it to avoid confusion.

For example, if you choose "valve fails open" this phrase is understood to cover three conditions: (1) the valve fails open, (2) the valve fails to remain closed, and (3) the valve fails to close. As you can see, sticking with the same convention makes your analysis more clear and simplified. Most of the conventions in general use appear in the Glossary at the end of this book under "Conventions."

4.2.4 Specialized workstations and software

If you have a significantly large population of equipment to manage such as an automobile assembly line, a major petrochemical plant, an oil refinery, a paper mill, or a large aerospace or computer manufacturing facility, there are specialized workstations and CMMS systems commercially available to help manage the data. There are also computer-based RCM tools that will store your data and monitor the performance of your equipment. All equipment I.D.s can be loaded into these software programs, and they will schedule the work, personnel-load the required PM tasks, monitor equipment performance criteria, track key performance indicators (KPIs), and identify when certain equipment needs to be addressed as a result of warning or alert limits that monitor preestablished equipment parameters.

Many facilities already have various databases for ready access and retrievability of all facets of their data. For these facilities this is a vital administrative requirement that ensures real-time information is readily accessible and retrievable from all requisite databases to be used for scheduling, monitoring, trending, documenting, or analyzing reliability data. There are several software programs that are commercially available to help manage this work. For smaller and midsized facilities, an Excel spreadsheet is an excellent tool and will work just fine.

Oftentimes, the management of a facility with internal programming resources available will opt to develop their own RCM software that is customized for their specific needs. This is a very efficient way to establish an RCM database because any changes in format can be accomplished internally with in-house personnel, and reliance on outside consultants is not needed. The decision to develop your own software or purchase software off the shelf is strictly a matter of cost and how much autonomy and control you wish to have over your own program.

4.2.5 The COFA spreadsheet versus the FMEA

The COFA can be completed either in its simplest format, which is an Excel spreadsheet, or if you prefer, you can embed it in a more formal software program. The logic remains the same. The COFA Worksheet will be one of the tools you will use when we commence the analysis implementation. The actual implementation process and how to complete the COFA using examples are thoroughly explained in Chapter 5.

If you study the differences between Figures 4.1 and 4.2 carefully, you will begin to understand why I created the COFA and you will see how simpler and more accurate it is than the customarily used FMEA. Looking at the FMEA in Figure 4.1, you will see that you have to start with a system and then define functions for each individual subsystem. The systems and sub-

	A	B	C	D	E	F	G	H
1								
2								
3								
4								
5	SYSTEM BEING ANALYZED: ______________							
6	SUBSYSTEM BEING ANALYZED: ______________							
7								
8								
9	DEFINE ALL	DEFINE SUBSYSTEM	DEFINE THE DOMINANT	IDENTIFY THE COMPONENT	IS THE OCCURRENCE	DEFINE THE SYSTEM		DEFINE THE
10	SUBSYSTEM	FUNCTIONAL FAILURES	COMPONENT FAILURE MODES	I.D.'s ASSOCIATED WITH THE	OF THE FAILURE	EFFECT FOR EACH	DEFINE THE	COMPONENT
11	FUNCTIONS	FOR EACH FUNCTION	FOR EACH FUNCTIONAL FAILURE	DOMINANT FAILURE MODES	EVIDENT?	FAILURE MODE	PLANT EFFECTS	CLASSIFICATION
12								
13								
14								
15								
16								
17								
18								
19								
20								
21								
22								
23								
24								
25								
26								
27								
28								
29	ANALYSIS PERFORMED BY: __________/__________/__________							
30	DATE: __________							
31								
32								

Figure 4.1 Failure Modes and Effects Analysis (FMEA) Worksheet.

	A	B	C	D	E	F	G	H
1								
2								
3								
4								
5								
6								
7							DESCRIBE THE "CONSEQUENCE	
8					IS THE OCCURRENCE		OF FAILURE" BASED ON THE	
9				DESCRIBE THE DOMINANT	OF THE FAILURE MODE		"ASSET RELIABILITY"	
10		DESCRIBE ALL	DESCRIBE THE WAYS	COMPONENT FAILURE	EVIDENT?	DESCRIBE THE	CRITERIA YOU SPECIFIED	DEFINE THE
11	COMPONENT I.D. &	FUNCTIONS OF THE	THAT EACH FUNCTION	MODES FOR EACH	(Determined by the	SYSTEM EFFECT FOR	(Determined by the COFA Logic	COMPONENT
12	DESCRIPTION	COMPONENT	CAN FAIL	FUNCTIONAL FAILURE	COFA Logic Tree)	EACH FAILURE MODE	Tree and the PC & ES Guidelines)	CLASSIFICATION
13								
14								
15								
16								
17								
18								
19								
20								
21								
22								
23								
24								
25								
26								
27								
28								
29	ANALYSIS PERFORMED BY: ____/____/____							
30	DATE: ____							
31								
32								

Figure 4.2 Consequence of Failure Analysis (COFA) Worksheet.

systems must be compartmentalized, and you *are required* to establish the cumbersome process of identifying boundaries and hundreds of interface points for each system.

In column A of the FMEA, all of the subsystem functions for each system must be defined. This is not a straightforward process because there is no practical handle on capturing all of the functions by system and subsystem. It entails a lot of research, and there is no assurance that you have captured all of them. Operating manuals, training guides, safety analysis reports, design documents, and the like are just some of the resources you will need. In column B, you will need to identify the functional failures of the subsystems.

In column C of the FMEA, you need to define the dominant component failure modes for the functional failures. Then in column D of the FMEA, you must identify the applicable equipment I.D.s associated with those dominant component failure modes defined in column C. Again, there are no practical checks and balances for identifying all of the components that might result in the previously specified functional failures. As I mentioned in Chapter 3, after completing the FMEA, it is not uncommon—even for seasoned RCM analysts—to end up with stray or leftover components that were not accounted for when identifying functions by system.

It is also inevitable that the same component I.D.s will be repetitively identified for different system/subsystem functions. Until now, the FMEA (or FMECA) format was the only option to use for a classical RCM program. That is why I created the COFA. The COFA format has all of the elements of the FMEA, but it includes additional attributes that make it all-encompassing, so that it is much easier to use and understand and thus provides for a much more accurate analysis.

Let's look at the COFA in Figure 4.2. System and subsystem boundaries and interfaces are *not required* because the analysis is performed at the component level. Note that the components are identified in the COFA in column A right at the start. The functions and functional failures are then defined at the component level for each component. Unlike the FMEA, with the COFA you do not need to figure out and research which components are responsible for the failures of the function at the system and subsystem levels because you already identified them at the beginning.

It is not a difficult task to identify the functions of a given component. When you already know what the component is, it is much easier to identify what the functions of that component are, rather than starting at a more abstract level of defining functions at the system and subsystem level and then figuring out how to back-fit the analysis by identifying what component meets that functional failure criterion.

As previously noted, any questions regarding the analysis are always at the component level anyway, which is another reason I developed the COFA. Even though there will be more individual component functions than system functions, the simplicity, clarity, and accuracy gained and the benefit of eliminating the requirement for establishing boundaries and interfaces are worth it.

As I mentioned in Chapter 3, I contend that specifying the functional failures is of questionable value because the functional failures really become the exact opposite of the functions. For instance, if the function is to "supply water to the injection header," then the functional failure is written as "fails to provide water to the injection header." Since the functional failure becomes the exact opposite of the function, it does not add any significant value. After defining the function, it is not that difficult to go directly to column C of the COFA and define the "dominant failure modes for those functions." Nonetheless, I have included defining the functional failures for two reasons. It does add a very marginal value of clarity to the process, and SAE has included it in their standard. However, including verbiage for the functional failures does not add a great amount of significance to the analysis.

Similarly, defining the failure effects at the system level for each failure mode is not really a significant value-added part of the process because it is the consequence of failure at the plant level that is important. I have nevertheless included system effects as an informational element because I have found that this knowledge makes the identification of the plant effects a little bit more clear. Some RCM programs even specify the effects at the local level, which is of miniscule value.

In the COFA, I have explicitly identified, in column G, that the consequence of failure is based on the asset reliability criteria that you establish. Asset reliability consequences include the plant-level consequences and any other consequences you deem important such as an incident resulting in unfavorable publicity, or a legal concern, or an environmental concern. Selecting your asset reliability criteria is discussed in Chapter 5.

Chapter 5 also explains the logic that prevails in order to complete each column of the COFA Worksheet. Remember, there is nothing incorrect about using the FMEA, and if you prefer the FMEA format, that is an acceptable way to perform your analysis. The choice is yours.

4.2.6 The PM task worksheet

As we learned in Chapter 3, phase 2 of an RCM program is where we *specify* the preventive maintenance tasks for the equipment we *identified* in phase 1, since those were the components requiring a preventive maintenance strategy. Run-to-failure components do not have PMs. That takes us to the PM Task Worksheet (PMTW). Refer to Figure 4.3 for the format of an Excel PM Task Worksheet. At this time, it is shown only to introduce you to its format since it will be one of the tools you will be using when we commence the analysis implementation. The actual PM selection process and how to complete the PM Task Worksheet using examples are thoroughly explained in Chapter 6.

The PM Task Worksheet is where we enter the data for all components that were classified as being either *critical, potentially critical, commitment,* or *economic.* These components must have a preventive maintenance strategy with "applicable and effective" PM tasks specified. The only exception is for failure modes that *are not* credible. However, noncredible failure modes are extremely rare because virtually any type of failure can occur. Even if a failure mode has not occurred previously, that is not a valid justification for calling it noncredible. I would grant that a meteor strike would qualify as being noncredible and per-

	A	B	C	D	E	F	G
1							
2							
3							
4							
5	NOTE: The PM TASK WORKSHEET IS ONLY APPLICABLE FOR THOSE COMPONENTS CLASSIFIED AS BEING EITHER *CRITICAL*, *POTENTIALLY CRITICAL*, *COMMITMENT*, OR *ECONOMIC*.						
6							
7		WHAT WERE THE			DESCRIBE THE "APPLICABLE	DEFINE THE FREQUENCY	
8	COMPONENT I.D. &	CONSEQUENCES OF	DESCRIBE EACH DOMINANT	DESCRIBE THE CREDIBLE FAILURE	AND EFFECTIVE" PM TASK	AND INTERVAL FOR EACH	
9	DESCRIPTION	FAILURE?	COMPONENT FAILURE MODE	"CAUSE" FOR EACH DOMINANT	FOR EACH FAILURE "CAUSE"	PM TASK	IS A DESIGN CHANGE
10	(From the COFA)	(From the COFA)	(From the COFA)	FAILURE MODE	(From PM Task Logic Tree)	(From PM Task Logic Tree)	RECOMMENDED?
11							
12							
13							
14							
15							
16							
17							
18							
19							
20							
21							
22							
23							
24							
25							
26							
27							
28							
29	ANALYSIS PERFORMED BY: ____________ / ____________ / ____________						
30	DATE: ____________						
31							
32							

Figure 4.3 PM Task Worksheet.

haps other, similar types of postulated events. Anything less would most likely be considered credible.

Each failure mode has one or more PM tasks to address it. Therefore, a single component will often have a host of different types of PM activities associated with it. The PM Worksheet is where we introduce the piece parts, or the subassemblies, of the components that are the credible causes for failure. Causes of failure could include bearing failures, heat exchanger tube failures, motor winding failures, check valve hinge pin failures, pump seal failures, impeller shaft failures, and so on. Chapter 6 defines the different types of PM tasks and explains in detail each column of the PM Task Worksheet.

If you have a rather large and complex facility, you may wish to use some of the previously mentioned commercially available RCM software to store this information. For smaller and midsize facilities that prefer to keep their costs at a minimum, the Excel PM Task Worksheet contains all of the required data and attributes—although they may not be as sophisticated as those embedded in a chic software package, they are equally as robust.

4.2.7 The economic evaluation worksheet

As we now have learned, economic components have no safety or operational consequences resulting from their failure. They result only in a monetary cost of labor and/or materials. The question often arises "What dollar-value threshold should be used as a break-even point for whether or not to perform a PM?" This is a subjective decision, and I have included some guidance for your decision process. Basically, it boils down to the economic cost of failure versus the cost for performing a PM to prevent its failure. Oftentimes the cost of a PM is far greater than letting an economic component fail. Remember, the difference between an *economic* component and a *run-to-failure* component is the cost involved. There are no other safety or operational concerns.

To allow for a break-even equivalency, everything should be calculated on an annualized basis. In a simple typical situation, for example, if the PM task needs to be accomplished two times per year to prevent the component from failing, and administrative costs average $500 each time the PM is accomplished, this would equal $1000 per year. Assume the component failed on an average of once per year, and it cost $1000 for repairs or replacement. That would result in a total annual cost of $1000 for the repairs or replacement plus $500 in CM administrative costs (assuming the CM admin costs are equivalent to the PM admin costs), for a total cost of $1500. Therefore, it would cost $1000 per year to maintain it, but it would cost $1500 per year to let it run to failure. So a PM strategy would be appropriate in this case.

Obviously, this was a very simple example. Real-life instances are much more complicated and must take into account many other variables, such as: Will a downpower be required each time the PM is performed? How long will the downpower last? How much will the downpower cost for that length of time? Is the component easily accessible to work on? Will scaffolding be required? Does it require an outage for clearance to work on the component? Are there warehouse stocking issues? Other warehouse costs? Are parts readily available? If the component is RTF and it fails on the back shift, will there be sufficient personnel to effect repairs? Can repairs wait till the next day? And so on . . .

Figure 4.4 is an Excel Economic Evaluation Worksheet to help identify a break-even point for your evaluation. Many times a plant wishes to choose an arbitrary dollar value, such as $500, so that if the component replacement cost is less than $500, it is not considered for preventive maintenance (again, in the absence of any safety or operability considerations). It is also common to calculate the administrative cost of performing a routine PM task. This cost would include the time needed for a scheduler to plan and schedule the task, the cost for processing the paperwork (which could be quite significant), the cost for operations support time, and so on. If these cumulative administrative costs are more than the cost of replacing the component, it may be prudent to consider the component as RTF (again, in the absence of any safety or operational concerns).

Some tasks, such as routine lubrication and oil changeouts, are inherently justified and prudent to perform, and they do not need any further economic evaluation.

Chapter Four

PM COST

(A) LABOR WORKER-HOUR (WH) COSTS:
    1) MAINTENANCE WHs = _____ @ $_____ / WH = $_____
    2) OPERATIONS WHs = _____ @ $_____ / WH = $_____
    3) OTHER WHs = _____ @ $_____ / WH = $_____
    Total Labor Cost / Yr = $_____ × _____ Number of PMs / Yr = $_____

(B) PM MATERIAL COSTS:
    Material Cost / PM = $_____ × _____ Number of PMs / Yr = $_____

(C) MISCELLANEOUS COSTS:
    1) Rental equipment cost = $_____ × _____ Number of PMs / Yr = $_____
    2) Other misc costs = $_____ × _____ Number of PMs / Yr = $_____
    Total Misc Cost / Yr = $_____ × _____ Number of PMs / Yr = $_____

TOTAL PM COST PER YEAR = A + B + C = $_____ /
    NOTE: If the PM is accomplished every 2 years, number of of PMs per year would be 0.5.

RTF COST

(A) LABOR WORKER-HOUR (WH) COSTS:
    1) MAINTENANCE WHs = _____ @ $_____ / WH = $_____
    2) OPERATIONS WHs = _____ @ $_____ / WH = $_____
    3) OTHER WHs = _____ @ $_____ / WH = $_____
    Total Labor Cost / Yr = $_____ × _____ Number of failures / Yr = $_____

(B) MATERIAL COSTS:
    Material Cost / failure = $_____ × _____ Number of failures / Yr = $_____

(C) MISCELLANEOUS COSTS:
    1) Rental equipment cost = $_____ × _____ Number of failures / Yr = $_____
    2) Other misc costs = $_____ × _____ Number of failures / Yr = $_____
    Total Misc Cost / Yr = $_____ × _____ Number of failures / Yr = $_____

TOTAL RTF COST PER YEAR = A + B + C = $_____ /
    NOTE: If the component fails every 3 years, on average, number of failures per year would be 0.33.

COST COMPARISON

If the PM costs per year are greater than the RTF costs per year, the component should be RTF (in the absence of any safety or operational concerns).

If the PM costs per year are less than the RTF costs per year, the component should be part of the preventive maintenance program.

ANALYSIS PERFORMED BY: _____ / _____ / _____
DATE: _____

Figure 4.4 Economic Evaluation Worksheet.

4.3 Chapter Summary

- Preparation for implementing your analysis includes selecting an RCM point of contact and selecting qualified representatives from operations, engineering, maintenance, and the various crafts.
- An alphanumeric database of all plant equipment is required.
- Informational resources to ascertain the functions of all plant equipment and the consequences of their failure are required.
- A convention for describing failure modes needs to be established.
- For the majority of facilities, an Excel spreadsheet is all that is required, and it will become the tool of choice for completing the RCM analysis.
- The COFA Worksheet, the PM Task Worksheet, and the Economic Evaluation Worksheet are all displayed in an Excel format.
- More sophisticated workstations, CMMS systems, and software are commercially available for larger and more complex facilities.
- The development of software by an internal, in-house IT department is also an option.

This chapter touched very briefly on the kinds of spreadsheet tools you need to proceed with the analysis and store the analysis data. Everything we learned about the concepts and principles of RCM and the tools we need have brought us to this point. Now let's look at actually commencing a classical RCM analysis.

Chapter

5

RCM Made Simple: The Implementation Process

This chapter guides you through the RCM implementation process. What we have learned thus far has been the building blocks for the implementation process, and as in previous chapters, I will build the implementation process in a sequence beginning with the essential elements. These elements include the requirement to identify your asset reliability criteria, which are the assets you want to preserve. Then you will need to identify the failure modes you want to prevent in order to avoid any unwanted consequences that can threaten the preservation of those assets. This is accomplished via the COFA, which identifies the population of equipment that must be included in a preventive maintenance strategy. Specifying the different types of PM tasks, establishing a PM task strategy, and completing the PM Task Worksheet are discussed in Chapter 6.

It has been my experience that the implementation process can be made very painful or it can be made virtually painless and just as robust, if not more so, than the painful method. Most RCM books discuss establishing a network of facilitators, a stringent parliamentary process for conducting meetings, and adherence to special training rigidities. These regimented requirements are not incorrect; they are just not needed. I have developed this process in such a way that the only regimentation necessary is to adhere to the descriptive interpretation explained here as you complete each step of the COFA Worksheet. The only basic requirements for the analysis are a complete database of all plant equipment; a qualified RCM point of contact; qualified representatives from the operations, engineering, and maintenance departments; representatives from the various crafts; informational resource data; and the cumulative knowledge and experience of the representatives for identifying all of the different functions of the equipment, the failure modes of the equipment, and the consequences of those failure modes.

I recommend that the representatives rotate from the different departments depending on the equipment and systems being analyzed. For example, the individuals most familiar with the electrical distribution system components are probably not the most qualified to analyze the feedwater system components, and vice versa—the mechanical folks are probably not the best individuals to analyze the electrical distribution system components. However, the individual acting as the RCM single point of contact should remain permanent.

There are no special training needs for these individuals other than their knowledge, expertise, and willingness to work together. The logic I have created is self-explanatory. The team members can determine for themselves how to conduct a meeting, what to say, when to speak, and what information needs to be documented. The representatives selected should be quite capable of determining the method for deciding which system components are the first to be analyzed. Eventually every component will be analyzed, so the order in which they are analyzed is not of paramount importance. Some of the components thought to be the least important might actually be found to be the most critical ones.

Taking an approach to RCM that is too sophomoric and parochial is counterproductive and unnecessarily costly, in my opinion. Too much emphasis on the mundane can result in a waning of interest in the whole RCM process. Approaching RCM in this way has contributed to what made the process so difficult in the past and why its implementation failure rate has been so high. RCM was not intended to be that difficult to implement, and it shouldn't be. As I mentioned in Chapter 2, the decisions required for successfully implementing an RCM program are best made by knowledgeable in-house individuals or by augmented help, working on an individually contracted basis, under in-house supervision.

The implementation starting point is to define your asset reliability criteria.

5.1 Define Your Asset Reliability Strategy

RCM is the *driver* for a corporate reliability strategy. As we learned in Chapter 1, your number one objective is preserving your corporate assets via a premier RCM program that ensures the highest levels of safety and reliability for your facility. Any progression of corporate mishaps such as employees or members of the general public incurring serious injuries, or having enforcement inspection teams or an OSHA committee descend upon your facility, will generate enough unwanted corporate visibility and uninvited public attention that the resulting negative impact would far outweigh the cost of implementing a prudent reliability program.

The first step in implementing your RCM program is to define precisely and accurately what asset reliability criteria your senior management wishes to preserve. Defining your asset reliability criteria basically involves identifying all the unwanted consequences of failure that can occur in your facility and that must be prevented.

To define these criteria, you should take into account the following concerns: ensuring personnel and plant safety; preventing, or at the very least minimizing, the exposure to any unplanned production delays; unplanned facility shutdowns, power reductions, and production interruptions; the loss of generation or production capacity; preventing regulatory or environmental issues from attracting unwelcome publicity or litigation; and so on. Defining these criteria is your first order of business and it is strongly advised that you submit a formal internal memorandum for signatures from all applicable individuals in authority for agreement on what you choose.

It is perfectly acceptable to place qualifying conditions on the criteria you select—for example, a production or assembly-line interruption with a duration of no more than 15 minutes, or 30 minutes, or 1 hour, or a power reduction of no more than 1 percent, 5 percent, or 10 percent. Other things to consider are:

- Can the plant manage the failure?
- Can compensatory measures be taken immediately and are they acceptable?
- Will there be enough time to mitigate any unwanted consequences?

Any qualifying conditions you choose should also have the buy-in of senior management. Once established, they will become the bedrock of your reliability program. For example, a manufacturing or production facility might incorporate these criteria: no personnel or safety concerns, no plant trips, no power reductions greater than 5 percent, no regulatory issues resulting in enforcement or unwanted oversight, and no control room actions or interruptions that cannot be mitigated within 10 minutes. Figure 5.1 lists some of the many possible criteria to consider.

While generic catchphrases such as "resource optimization" and "plant optimization" are worthy, they are not sufficiently specific. You need to be very specific about what you mean by such terms. The typical asset reliability criteria listed in Figure 5.1 are very specific, reflecting your need to be as specific in your criteria in order to achieve reliability success.

In Figure 5.1, you will note that it is *mandatory* to include the first two criteria, which are *personnel, public, and plant safety* criteria. The others are all *operability* criteria. Of course, you may add to this list or omit some of the suggested criteria (except for the first two safety criteria). But once you and your senior management have decided upon and defined exactly what criteria you will use, that becomes the focus of your RCM analysis. *Economic* criteria are separate from *safety* and *operability* criteria and are discussed in conjunction with Figure 5.2b.

Component functions, functional failures, and the consequences of those failures *that can result in triggering one or more of your selected asset reliability criteria* will result in a component classification of either *critical* or *potentially critical*. As

Safety Concerns:
- No personnel safety or public safety concerns (*mandatory*)
- No plant or facility safety concerns (*mandatory*)

Operability Concerns:
- No unplanned plant or facility shutdowns
- No power reductions, downpowers, or production interruptions

    Qualifying conditions could include:
    - Greater than X percent
    - Longer than X minutes or X hours
- No power transients
- No technical specification violations or limiting conditions for operation (nuclear related)
- No minimum equipment list violations or limiting conditions for flight (airline related)
- No regulatory consequences or enforcement actions or vulnerability to such actions
- No environmental impact, consequences, or issues
- No litigation impact, consequences, or issues
- No nonroutine or abnormal control room evolutions

    Qualifying conditions could include:
    - Evolutions that cannot be mitigated within X minutes
- No operator required workarounds

    Qualifying conditions could include:
    - Workarounds for more than X hours or more than X days
    - Unless compensatory measures can be taken immediately
- No unplanned or inadvertent actuations of emergency systems
- No significant customer concerns or issues

Figure 5.1 Typical asset reliability criteria.

mentioned, *economic* criteria are separate from *safety* and *operability* criteria and are discussed later in this chapter. *Commitment* components are determined by the specific commitment requirements you have agreed to with your respective regulators. Therefore, your PM program will consist of components classified as one of the following:

- Critical safety or operability
- Potentially critical safety or operability
- Commitment
- Economic

Any component with one of these classifications *should* have a preventive maintenance strategy to prevent the consequence of its failure, or a design change should be implemented if an applicable and effective PM task cannot be specified.

I suggest that you code each criterion with a number or an abbreviation—for example, C-PerS, C-PubS, or C-PlaS for a critical component that has a consequence of personnel safety, public safety, or plant safety. It could also be abbreviated PC-EI or PC-PSD for a potentially critical component with a consequence of an environmental impact or a plant shutdown. The reason for doing this is that at some later time you may wish to sort all of the components according to, for example, those whose failure could result in an unwanted plant shutdown, or a power reduction, or a regulatory consequence. It is a good data resource that adds clarity and precision to the critical or potentially critical classification.

5.2 Understanding the RCM COFA Logic Tree, the Potentially Critical Guideline, and the Economically Significant Guideline

Before we get into the COFA Worksheet, you need to have some understanding of the RCM COFA Logic Tree, the Potentially Critical Guideline, and the Economically Significant Guideline. This is truly *RCM Implementation Made Simple*. You will find

RCM Made Simple: The Implementation Process

that I have taken a complex logic process and reduced it to its basic elements while enhancing its thoroughness, accuracy, and clarity. This also builds on what we learned in Chapter 3 about critical, potentially critical, commitment, and economic component classifications and the asset reliability criteria you selected. Refer to Figure 5.2a for the COFA Logic Tree and Figure 5.2b for the Potentially Critical Guideline and Economically Significant Guideline. I will walk you through each step of the entire logic process in Figures 5.2a and 5.2b, explaining each decision you will make. The logic begins by identifying critical components, then potentially critical components, and then commitment and economic components.

Note that the logic is created such that in order to be classified as economic, a component must first go through the COFA Logic Tree and the Potentially Critical Guideline to ensure it is not just an economic component, because it could also be a critical or potentially critical component. Most RCM programs will immediately jump right to an economic conclusion if the component

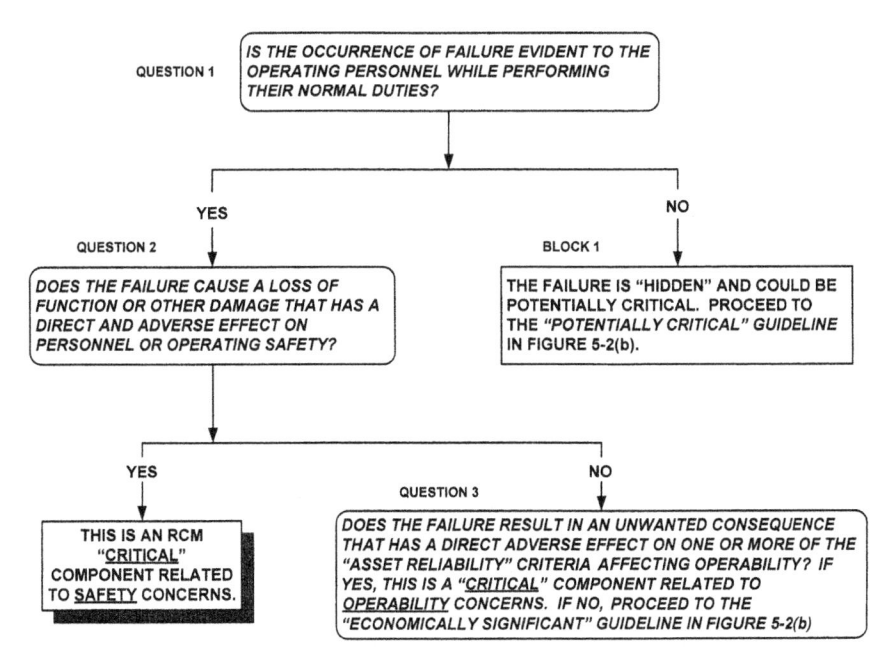

Figure 5.2(a) RCM COFA Logic Tree.

Potentially Critical Guideline

Can the component failure, in combination with an additional failure or initiating event, or over time, result in an unwanted consequence that has a direct adverse effect on one or more of the asset reliability criteria?

If *yes*, this is a *potentially critical* component. It could be "potentially critical" for *safety* or for *operability* concerns depending on its consequence of failure.

If *no*, is the component associated with a commitment? If it is, this is a *commitment* component. If it is not associated with a commitment, proceed to the Economically Significant Guideline below.

Economically Significant Guideline

- Will the component failure result in a high cost of restoration?
- Will the component failure result in a high cost of related corrective maintenance (CM) activity?
- Will the component failure result in significant downtime?
- Will the component failure result in a long lead time for replacement parts? Are the parts obsolete or in short supply?

If *yes* to any of the above, and a PM is further justified by the Economic Evaluation Worksheet (Figure 4.4), this is an *economic* component.

If *no* to all of the above, this is a *run-to-failure* (RTF) component.

Figure 5.2(*b*)

has a high number of corrective maintenance orders or a high cost of restoration, for example. That is not acceptable because it is not necessarily correct. In many cases, the component will also be either critical or potentially critical, which is much more important than economic. As we learned in Chapter 3, the component classification always defaults to the highest level. This possibility would otherwise go undetected. Immediately jumping to an economic conclusion would never expose the component as also being critical or potentially critical.

To put all of this into perspective, much of the difficulty and confusion with existing traditional renditions of RCM, especially for the RCM novice, arises with the disjointed decision logic process. Any real cohesiveness was missing from the logic process. What I have done is to simplify the entire logic process by developing the concept of potentially critical components and very specifically and very clearly integrating all of the decision logic into one all-inclusive, self-contained process. I have designed the COFA Worksheet so that it embraces the following simplicity:

- The entire analysis process begins at the component level.
- Up front, you have already established your asset reliability criteria, which constitute your standard for determining safety and operability concerns.
- The COFA Logic Tree then identifies critical components through the RCM decision process if it affects any of the asset reliability criteria you established.
- In case the component is not deemed critical by the COFA Logic Tree, the Potentially Critical Guideline was embedded in the decision logic process to determine whether a hidden failure could affect your asset reliability criteria, which would make the component potentially critical.
- If the Potentially Critical Guideline decision logic did not find the component to be potentially critical or have a commitment associated with it, the Economically Significant Guideline was embedded in the decision logic to determine whether the component failure could result in an economic concern.

Thus, the entire RCM decision logic has been integrated into one very simple process. The COFA Worksheet that we discussed in Chapter 4 refers you to the logic required to determine the ultimate *consequence of failure*. Since all of the decision logic has been sequentially integrated within the RCM COFA Logic Tree, the Potentially Critical Guideline, and the Economically Significant Guideline, it is only prudent to review them before we begin to complete the COFA Worksheet.

The following is the descriptive logic for each question in the RCM COFA Logic Tree.

Question 1: Is the occurrence of failure evident to the operating personnel while performing their normal duties?"

There is a reason that this is the very first question asked: it is to distinguish between evident failures and hidden failures, and that is where the RCM logic begins. This question must be answered for *each* failure mode. A component usually has several failure modes. The failure modes are those that are dominant, or most likely to occur. They do not include implausible or unrealistic failure modes. Failure modes are the types of failure or the ways in which a component can fail. Failure modes could be, for example, "valve fails closed," "valve fails open," "switch fails to actuate," "switch actuates prematurely," "pump fails to start," and so on. Nonplausible failure modes are similar to noncredible failure causes, which we discussed in Chapter 4. They are the extreme rarity. (*Note:* The failure mode does not include the *reason* or the *cause* for the component failure at this point in the analysis. The credible failure *causes* are analyzed during the PM task evaluation phase, which is covered in detail in Chapter 6.)

The failure of the component could be made evident by a control room alarm, an indicator light, a valve position indicator, or any other control room announcement that monitors the operation of that specific component. This provides an indication that the component has failed and an urgent situation has arisen. For example, if the failure of a particular pump occurs, then the plant must reduce power immediately. Of course, the failure could also become evident via the unwanted event that occurs simultaneously with the failure—in other words, there is no time to take immediate action because the consequence has already occurred. For example, as soon as a conveyor failure occurs, the assembly line shuts down. Any failure mode that is not evident is considered hidden.

As I mentioned in Chapter 3, every facility is not like the control room of a nuclear power plant or the cockpit of a jet aircraft, where almost all equipment and systems are continuously monitored. If your facility does not continuously monitor the majority of your equipment in a control room, it is acceptable to include the failure as "being evident" if monitoring of the equipment is accomplished during routinely performed operator rounds. Practically all facilities have an individual(s) who continuously monitors plant equipment and other parameters throughout a 24-hour period. To receive credit for an event being evident, the operator rounds cannot involve some loosely interpreted set of verbal instructions that can arbitrarily change at will. The operator rounds must be formalized and defined within specific applicable operating procedures. We will examine both *yes* and *no* answers to Question 1, starting with a *yes* answer, which brings us to the next question.

Question 2: Does the failure cause a loss of function or other damage that has a direct and adverse effect on personnel or operating safety?

This question also must be asked for *each* failure mode. This is the most serious of all consequences. A *yes* answer to Question 2 means the *single failure* can result in a direct adverse effect on personnel or plant safety. A single failure of the component can immediately result in the possibility that an employee or a member of the public could be seriously injured or that the failure could lead to a fatality or a serious safety concern for the plant. An example is the failure of the O-ring seal on the *Challenger* space shuttle. A *yes* answer could also mean that there is the possibility of a very serious plant safety concern such that the component failure might result in a fire or explosion in the plant (which would also be a personnel safety issue). This would be a critical component related to safety concerns. (*Note:* It is understood that all consequences of failure, whether at the system level or the plant level, *imply that there is no existing preventive maintenance strategy for preventing the failure.* For example, if you have an existing PM that performs a task, having that PM must not be considered in the analysis to determine the consequence of failure. The existing PM task may indeed be applicable and effective, but it does not negate your requirement to define the criticality of the component by assuming the component will fail without crediting any existing preventive maintenance activities.)

A *no* answer to Question 2 brings us to Question 3.

Question 3: Does the failure result in an unwanted consequence that has a direct adverse effect on one or more of the asset reliability criteria affecting operability?

A *yes* answer to Question 3 determines this to be a critical component related to operability concerns. We know the failure was evident, and it was not determined to be a safety concern because we answered *no* to Question 2, and all of the remaining asset reliability criteria (except for the first two) are operability concerns. If it is not an operability concern, it could still be an economic concern.

A *no* answer to Question 3 directs us to proceed to the Economically Significant Guideline in Figure 5.2*b*. A *yes* answer to any of the economic questions determines the component to be classified as economic. A *no* answer to the Economically Significant Guideline determines the component to be run-to-failure.

Now let's look at a *no* answer to Question 1. That brings us to Block 1.

Block 1: The failure is hidden and could be potentially critical. Proceed to the Potentially Critical Guideline.

We then proceed to the Potentially Critical Guideline in Figure 5.2*b* to determine whether the component failure, *in combination with an additional failure or initiating event,* or over time, can result in an unwanted consequence that has a direct adverse effect on one or more of the asset reliability criteria, including the first two safety criteria.

A *yes* answer in the Potentially Critical Guideline determines the component to be potentially critical. An example of a potentially critical safety concern is similar to the triple-redundant (but hidden) turbine overspeed vulnerability we examined in Chapter 3, whereby the plant could sustain significant damage, but even worse, someone could be killed.

Note the distinction between Question 2 and the Potentially Critical Guideline in regard to safety. A *yes* answer to Question 2 determines that there is a direct adverse safety consequence that occurs immediately. A *yes* answer to the Potentially Critical Guideline affecting either of the first two asset reliability safety criteria also determines that there is a safety consequence, but it is not direct and immediate because it is hidden. Therefore, the most critical consequences of all will be those resulting in a *yes* answer to Question 2.

Components whose failure results in a *yes* answer to Question 2 or a *yes* answer in the Potentially Critical Guideline for either of the first two asset reliability criteria should be looked at for a possible *design change* in addition to any preventive maintenance strategies that may be implemented.

A failure resulting in a consequence to any of the other asset reliability criteria (except the first two safety criteria) would represent a potentially critical component related to an operability concern. A typical example is the hidden failure of a particular motor that could result in an unplanned interruption of production with the additional failure of a second component.

A *no* answer in the Potentially Critical Guideline asks whether the component has a commitment associated with it. If it has an associated commitment, it is classified as a commitment component. If it does not have an associated commitment, we proceed to the Economically Significant Guideline to determine whether the component should be classified as economic or run-to-failure.

Figure 5.3 illustrates what I call the *RCM filter*. This is a simplified way of looking at the logic we have just learned about, showing how it ensures that nothing gets through the funnel without being filtered for its importance.

As you can see, the logic I have created provides for *all* components to initially pass through the first filter, which is the COFA Logic Tree. All critical components are determined by this first filter. Those components making it through the first filter then pass through the second filter, which is the Potentially Critical Guideline. All potentially critical components are determined by this second filter. Those components making it through the first two filters then must pass through the commitment filter, which is included within the Potentially Critical Guideline. Those components making it through the first three filters then pass through the fourth filter, which is the Economically Significant Guideline. All purely economic components are determined by this fourth filter.

All of the components that were filtered constitute the technical basis for the preventive maintenance program. If a component makes it through the funnel without being caught in any of the filters, it is a run-to-failure component. This is truly *RCM Implementation Made Simple*.

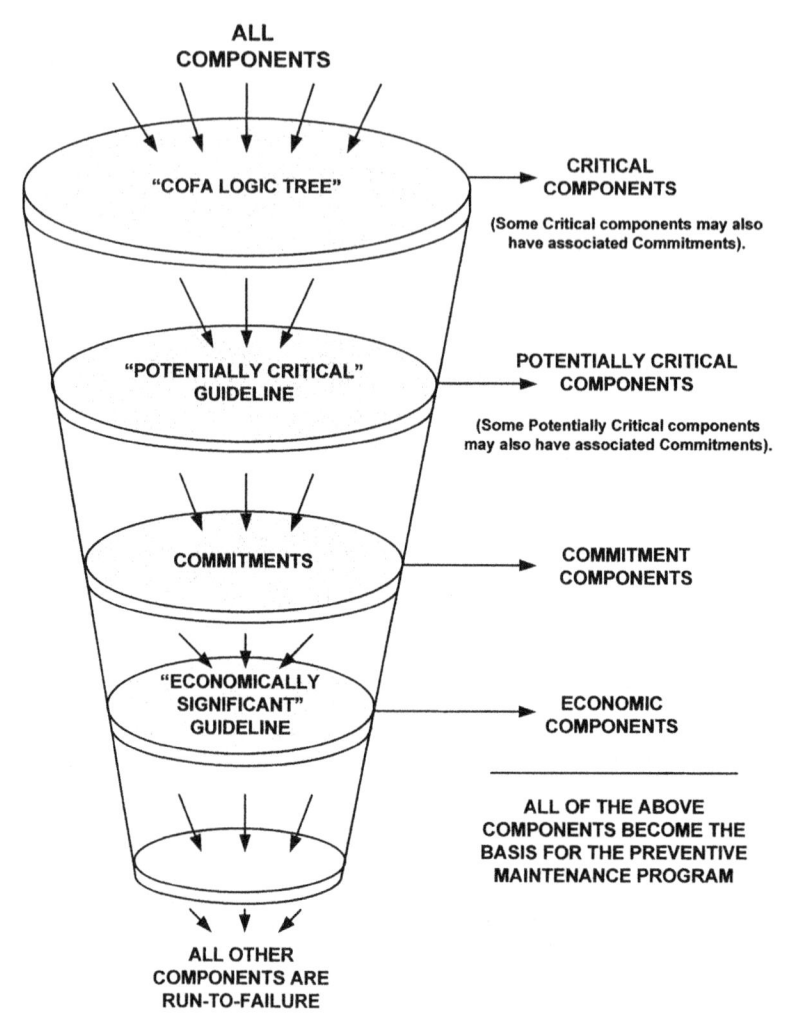

Figure 5.3 RCM filter.

5.3 Completing the COFA Worksheet in Conjunction with the COFA Logic Tree, the Potentially Critical Guideline, and the Economically Significant Guideline

With an understanding of the COFA Logic Tree, the Potentially Critical Guideline, and the Economically Significant Guideline, we can begin our analysis on the COFA Worksheet. For simplicity, the COFA Worksheet integrates the COFA Logic Tree and the two guidelines.

In this example for completing the COFA Worksheet, we assume that we are analyzing a two-way isolation valve, equipment I.D. GHI, that provides a flow path for chilled water for cooling heat exchangers X and Z. Figures 5.4a through 5.4g show how you enter the data for each column of the worksheet as we proceed with our analysis. I could have entered all of the information at once, but doing so step-by-step makes it easier to follow and understand.

5.3.1 Describe the component functions

We begin by entering the component I.D. in column A of the COFA Worksheet, and in column B we describe *all* of the functions of the component. Refer to Figure 5.4a. There will be several functions for each component. These include all of the normal operating functions *and* any emergency functions of the equipment, such as those design functions that come into being as the result of the loss of on-site power, flooding concerns dur-

COFA Column A	COFA Column B
Component I.D. and Description	**Describe All Functions of the Component**
1. Valve GHI 2-way isolation vlv	1.a. Provide a flow path for cooling heat exchanger X when pump Y is operating.
	1.b. Provide for isolation of the cooling flow path to heat exchanger X when pump Y is not operating.
	1.c. Provide emergency cooling flow to backup heat exchanger Z during a main condenser malfunction.

Figure 5.4(a) Consequence of Failure Analysis (COFA) Worksheet.

ing heavy periods of rain or thunderstorms, the loss of cooling capability, the occurrence of a fire, and the like. The functions describe what the component must accomplish. The functions are the explanation for why the component is installed, and preserving these functions is the main objective of the maintenance program. The purpose for defining functions as part of the analysis is to enable the emergence of the specific failure modes and their respective consequences of failure.

For example, one function of two-way valve GHI is "to provide a flow path for cooling heat exchanger X while pump Y is operating." Another function for the same valve is "to provide for isolation of the cooling flow to heat exchanger X when pump Y is not operating." Another function for valve GHI is "to provide an emergency cooling flow path to backup heat exchanger Z in the event the main condenser malfunctions."

The function must be written such that there is a clear definition of the conditions that constitute a functional failure. Unfortunately, once again, the alleged RCM experts have overburdened this process element to the point that it has taken a feat of engineering wizardry to identify a simple function. I have seen RCM books and other RCM-related documents that specify a function to be written, for example, as follows: "provide a flow of 200 gal/min at 285 psig and a temperature of 87°F." Think about that for a minute. The functional failure would have to read "fails to provide a flow of 200 gal/min at 285 psig and a temperature of 87°F." What if the pump supplies only 199 gal/min at 284 psig and 86°F? Is that a failure of the function? Most likely, no—that is, unless you are talking about the loss of reactor coolant inventory at a nuclear power plant that is measured in ounces per day with precise, documented, regulatory requirements expressly specifying 200 gal/min at 284 psig and 87°F.

So in a nonnuclear, non-loss-of-reactor-coolant inventory situation, which is close to 99.9 percent of all other situations, what would the consequence of failure be if the pump output was only 199 gpm at 284 psig, and 86°F? Probably indiscernible. This is just another example how RCM careened out of control once it left the commercial aviation industry. It would be much more appropriate to write that function as: "provide the necessary flow needed to maintain the tank inventory at its nominal level."

In the absence of a regulatory requirement that specifically defines the operational parameters that the function must meet, the only functional definition you need to specify is a performance standard at a level determined by you, the owner of the facility, and the owner of your RCM program.

5.3.2 Describe the functional failures

In column C of the worksheet, you describe the ways that each function can fail. In what ways might each function be lost? Refer to Figure 5.4*b*. Typically, the functional failures are the exact opposite of the function. For example, if the function is "to provide a flow path for cooling heat exchanger X while pump Y is operating," the functional failure will be "fails to provide a flow path for cooling heat exchanger X while pump Y is operating."

As I stated in Chapter 4, since the functional failure typically becomes the exact opposite of the function, it does not add any

COFA Column A	COFA Column B	COFA Column C
Component I.D. and Description	**Describe All Functions of the Component**	**Describe the Ways Each Function Can Fail**
1. Valve GHI 2-way isolation vlv	1.a. Provide a flow path for cooling heat exchanger X when pump Y is operating.	1.a. Fails to provide a flow path for cooling heat exchanger X when pump Y is operating.
	1.b. Provide for isolation of the cooling flow path to heat exchanger X when pump Y is not operating.	1.b. Fails to provide for isolation of the cooling flow path to heat exchanger X when pump Y is not operating.
	1.c. Provide emergency cooling flow to backup heat exchanger Z during a main condenser malfunction.	1.c. Fails to provide emergency cooling flow to backup heat exchanger Z during a main condenser malfunction.

Figure 5.4(*b*) Consequence of Failure Analysis (COFA) Worksheet (*continued*)

significant value. After defining the function, it is not that difficult to go directly to column D and define the "dominant failure modes for those functions." Nonetheless, I have included defining the functional failures for two reasons: (1) it does add a very marginal value of clarity to the process, and (2) SAE has included it in their standard. However, including functional failures does not add a great amount of significance to the analysis.

5.3.3 Describe the dominant component failure modes for each functional failure

In column D, you describe the dominant component failure modes. Refer to Figure 5.4c. The failure modes are the different types of failure or the different ways a component can fail so that it fails to provide the function(s) specified. We include only plausible and realistic failure modes. For example, the dominant failure mode for the functional failure of "fails to provide a *flow path* for cooling heat exchanger X while pump Y is operating" would be "valve GHI fails *closed*." The failure mode for the functional failure of "fails to provide *isolation* for cooling heat exchanger X while pump Y is not operating" would be "valve GHI fails *open*." The failure mode for the functional failure of "fails to provide emergency cooling flow to backup heat exchanger Z during a main condenser malfunction" would be "valve GHI fails to *transfer* to its emergency position."

5.3.4 Is the occurrence of the failure mode evident?

In column E you identify whether the failure mode is evident. Refer to Figure 5.4d. This question comes directly from Question 1 of the COFA Logic Tree. It is the very first question asked and it determines whether the failure is evident or hidden. As we discussed in Section 5.2, to answer *yes* to this question, the failure mode must be evident when it occurs. It must be evident to operating personnel while they are performing their normal duties. This includes control room indication and monitoring, or it could include operator rounds if the rounds are routinely performed

COFA Column A **Component I.D. and Description**	COFA Column B **Describe All Functions of the Component**	COFA Column C **Describe the Ways Each Function Can Fail**	COFA Column D **Describe the Dominant Component Failure Modes for Each Functional Failure**
1. Valve GHI 2-way isolation vlv	1.a. Provide a flow path for cooling heat exchanger X when pump Y is operating.	1.a. Fails to provide a flow path for cooling heat exchanger X when pump Y is operating.	1.a. Valve fails closed.
	1.b. Provide for isolation of the cooling flow path to heat exchanger X when pump Y is not operating.	1.b. Fails to provide for isolation of the cooling flow path to heat exchanger X when pump Y is not operating.	1.b. Valve fails open.
	1.c. Provide emergency cooling flow to backup heat exchanger Z during a main condenser malfunction.	1.c. Fails to provide emergency cooling flow to backup heat exchanger Z during a main condenser malfunction.	1.c. Valve fails to transfer to the emergency position.

Figure 5.4(c) Consequence of Failure Analysis (COFA) Worksheet (*continued*).

and formally proceduralized. During the operation of your facility, sometimes the valve might be in the open position and other times it might be closed, depending on the operating parameters, but as long as the different valve positions are part of the routine monitoring and operating evolutions, the valve's failure will be evident. If any failure mode is not evident to the operating personnel, it is a hidden failure mode.

5.3.5 Describe the system effect for each failure mode

What are the failure effects at the system level? In column F we identify the system effects. Refer to Figure 5.4e. This is really an intermediate point of information because our ultimate goal is to identify the consequence of failure at the plant or facility level. System-level effects are included, nevertheless, as an informational element because they make the identification of the plant effects a little bit more clear. Oftentimes the failure does not result in any significant system effect, but there may be a regulatory requirement for its continuous operability and the failure may indeed result in a critical effect at the plant level.

Also keep in mind that all hidden failures have no system effect. By definition, "when a component is required to perform its function and the occurrence of the failure is *not* evident to the operating personnel, i.e., ... the immediate *overall operation of the system remains unaffected* in either the *normal* or the *demand* mode of operation, then the failure is defined as being *hidden*." One system effect that might be described even though the failure is hidden is the "loss of redundancy," and the hidden failure resulting in the loss of redundancy would most likely become potentially critical as a consequence of failure at the plant level.

Remember, too, that if a system is not normally operating, as it would not be in a safety system or a safety function, it is analyzed as though it is in its operating mode. Therefore, failure mode 1.c. is not hidden because it would be quite obvious if valve GHI failed to move to the emergency position when called upon to function.

System effect 1.a. delineates that "heat exchanger X will not provide cooling for pump Y, causing the pump to overheat." The

COFA Column B Describe All Functions of the Component	COFA Column C Describe the Ways Each Function Can Fail	COFA Column D Describe the Dominant Component Failure Modes for Each Functional Failure	COFA Column E Is the Occurrence of the Failure Evident?
1.a. Provide a flow path for cooling heat exchanger X when pump Y is operating.	1.a. Fails to provide a flow path for cooling heat exchanger X when pump Y is operating.	1.a. Valve fails closed.	Yes
1.b. Provide for isolation of the cooling flow path to heat exchanger X when pump Y is not operating.	1.b. Fails to provide for isolation of the cooling flow path to heat exchanger X when pump Y is not operating.	1.b. Valve fails open.	Yes
1.c. Provide emergency cooling flow to backup heat exchanger Z during a main condenser malfunction.	1.c. Fails to provide emergency cooling flow to backup heat exchanger Z during a main condenser malfunction.	1.c. Valve fails to transfer to the emergency position.	Yes

Figure 5.4(*d*) Consequence of Failure Analysis (COFA) Worksheet (*continued*).

COFA Column C	COFA Column D	COFA Column E	COFA Column F
Describe the Ways Each Function Can Fail	**Describe the Dominant Component Failure Modes for Each Functional Failure**	**Is the Occurrence of the Failure Evident?**	**Describe the System Effect for Each Failure Mode**
1.a. Fails to provide a flow path for cooling heat exchanger X when pump Y is operating.	1.a. Valve fails closed.	Yes	1.a. Heat exchanger X will not provide cooling for pump Y causing the pump to overheat.
1.b. Fails to provide for isolation of the cooling flow path to heat exchanger X when pump Y is not operating.	1.b. Valve fails open.	Yes	1.b. There is no system effect.
1.c. Fails to provide emergency cooling flow to backup heat exchanger Z during a main condenser malfunction.	1.c. Valve fails to transfer to the emergency position.	Yes	1.c. Heat exchanger Z will not provide cooling for emergency equipment.

Figure 5.4(e) Consequence of Failure Analysis (COFA) Worksheet (*continued*).

system effect of pump Y overheating will most likely result in a critical consequence of failure at the plant level. So as you can see, the purpose of identifying effects at the system level is only to offer a little more insight into the consequence of failure at the plant level.

5.3.6 Describe the consequence of failure based on the asset reliability criteria you selected

Column G is where we identify the consequence(s) of failure that can result in an unwanted impact on one or more of the asset reliability criteria you selected. Refer to Figure 5.1 for typical examples of these criteria. The consequence of failure is determined directly from the COFA Logic Tree, the Potentially Critical Guideline, and the Economically Significant Guideline. Refer to Figure 5.4f. All these unwanted consequences are at the plant or facility level, and they are identified for *each failure mode,* so there may be more than one consequence. For failure mode 1.a., the consequence is a power reduction greater than 25 percent for more than 10 hours. For failure mode 1.b., there is no plant consequence. For failure mode 1.c., the failure results in a plant shutdown. If there is more than one consequence, you need to include *all* of them. As noted earlier, at a later date you may wish to sort your asset reliability consequences according to the type of consequence, and it is therefore prudent to list all of them. For example, you may wish to sort all components whose failure can result in a plant shutdown.

5.3.7 Define the component classification

The component classifications for each consequence of failure must be identified in column H. Column H becomes quite important because any subsequent questions or challenges that arise in regard to what happens when valve GHI *fails open* will be documented. In this case, nothing happens, since failure mode 1.b resulted in an RTF designation. However, for failure mode 1.c., if valve GHI *fails to transfer,* a plant shutdown would occur, which is a critical consequence of failure.

COFA Column D	COFA Column E	COFA Column F	COFA Column G
Describe the Dominant Component Failure Modes for Each Functional Failure	**Is the Occurrence of the Failure Evident**	**Describe the System Effect for Each Failure Mode**	**Describe the Consequence of Failure Based on the Asset Reliability Criteria**
1.a. Valve fails closed.	Yes	1.a. The heat exchanger will not provide cooling for pump Y and it will overheat.	Failure of pump Y will result a power reduction of 25% for at least 10 hours.
1.b. Valve fails open.	Yes	1.b. There is no system effect.	There is no plant consequence.
1.c. Valve fails to transfer to the emergency position.	Yes	1.a. Heat exchanger Z will not provide cooling for emergency equipment.	Failure of backup heat exchanger Z will result a plant shutdown.

Figure 5.4(f) Consequence of Failure Analysis (COFA) Worksheet (*continued*).

Once again, you will have several different classifications for the same component, ranging from critical to run-to-failure. The final component classification always defaults to the highest level—that is, critical, potentially critical, commitment, or economic. Refer to Figure 5.4g. In this example, component GHI is classified as critical for two reasons: a power reduction and a plant shutdown. Even though a plant shutdown is more serious than a power reduction, I would include both consequences and classify valve GHI as *critical* due to a plant shutdown or a power reduction greater than 25 percent.

This completes the first part of the RCM journey and illustrates what RCM is all about: identifying the population of equipment whose failures can result in a negative impact on any of the asset reliability criteria you select. Your entire PM program will be focused on selecting applicable and effective PM tasks to prevent, eliminate, or mitigate failures of this population of equipment. It took a while to get to this point, and everything in the analysis thus far has been to drive us to identify these components so that their failures can be subjected to a preventive maintenance strategy.

This is phase 1 of the three phases of an RCM program, which you learned about in Chapter 3. Phase 2 includes the process of specifying the applicable and effective PM tasks for this population of equipment. I discuss the PM task strategies in Chapter 6.

5.4 RCM Serves as a Translation of the Design Objectives

RCM can be viewed, in a pure vision, as a *translation of the design functions of a component into the operating objectives of the facility via a reliability program.* Each component has a specific design function that must reliably meet the larger purpose, which is the operating objective of the plant. Each individual component must function properly within its respective system to ensure that the plant meets its ultimate design objective. The preventive maintenance program is the means by which this objective is achieved.

COFA Column E	COFA Column F	COFA Column G	COFA Column H
Is the Occurrence of the Failure Evident?	**Describe the System Effect for Each Failure Mode**	**Describe the Consequence of Failure Based on the Asset Reliability Criteria**	**Define the Component Classification**
Yes	1.a. The heat exchanger will not provide cooling for pump Y and it will overheat.	Failure of pump Y will result a power reduction of 25% for at least 10 hours.	Component is *critical* due to a power reduction.
Yes	1.b. There is no system effect.	There is no plant consequence.	Component is run-to-failure.
Yes	1.a. Heat exchanger Z will not provide cooling for emergency equipment.	Failure of backup heat exchanger Z will result a plant shutdown.	Component is *critical* due to a plant shutdown.

Figure 5.4(g) Consequence of Failure Analysis (COFA) Worksheet (*continued*).

5.5 Companion Equipment

An important check and balance for the RCM logic is to ensure that any companion equipment is also carefully analyzed. I regard *companion equipment* as those components associated with a critical or potentially critical component. Companion equipment could include an inlet or discharge check valve, a component providing an input signal, or a component that supports one of the functions of the critical or potentially critical component. It could even be a component as innocuous as a steam trap that supports moisture drainage for the critical or potentially critical component.

These companion equipment components should already have been independently analyzed themselves as either critical or potentially critical. However, when viewed as companion equipment, it is more likely that they will not inadvertently be overlooked. Since all components are analyzed regardless, it is a good check to be sure that those components associated with a critical or potentially critical component are also analyzed appropriately.

A typical example of a real-life situation again illustrates the importance of companion equipment. In a specific feedwater system, one of the feedwater pumps is driven by a steam turbine. The turbine and all associated components were identified as critical in the analysis. In a typical steam supply, there may be several hundred individual steam traps for controlled removal of accumulated moisture. Associated with the feedwater turbine, it was identified that failure of two of the rather innocuous steam traps could allow water to accumulate to the point of intrusion into the inlet to the feedwater turbine.

If this should occur, a slug of water will instantaneously slow the turbine down until it passes through, hopefully without damaging the impeller blades. However, slowing the turbine, even momentarily, engages the very sensitive turbine speed control function, causing the turbine to speed up to maintain its required RPM. Once the water slug no longer exists, the turbine now is in an overspeed condition, and when sensed by the speed control, it causes the turbine to shut down, also shutting down that supply of feedwater flow. Thus, these two steam traps constitute companion equipment. Not only were they classified as critical, but a design change was implemented to prevent the same situation from occurring in the future.

5.6 The SAE Standard: Document JA1011

The Society of Automotive Engineers (SAE) developed a standard that entitles a process to be called an RCM process. The main reason for this was that the RCM terminology was being applied to a multitude of PM program enhancements that had no technical basis or logic and were not systematically developed. They were a conglomeration of PM betterment efforts, or PM review efforts, or PM program updates, which were improperly described under the guise of RCM. The SAE wanted to ensure the isolation of these other, rather arbitrary, efforts from the more defined and thorough approach of specifically applying RCM logic. As a result, they issued a fundamental standard that had to be met in order to call the maintenance program process an RCM process.

The SAE Standard for RCM as delineated in the organization's Document JA1011 includes the following seven basic questions:

Question 1. What are the functions of the asset?

Question 2. What are the functional failures?

Question 3. What are the failure modes?

Question 4. What are the failure effects?

Question 5. What are the failure consequences?

Question 6. What are the PM tasks?

Question 7. What must be done if a PM task cannot be specified?

Every step in the COFA logic of *RCM Implementation Made Simple* is in concert with, and is in many aspects more advanced than, the SAE Standard for RCM. Five of the seven steps in the SAE Standard are covered by the COFA. The remaining two steps in the SAE Standard include specifying preventive maintenance tasks and identifying the default actions if an applicable and effective task cannot be identified. These two steps are covered in Chapter 6. The SAE Standard also mentions two "remaining issues," as they are referred to in that standard. These two remaining issues pertain to determining task intervals and establishing a continuous review of the RCM process. The task intervals are covered in Chapter 6, and the continuous RCM review process is covered in Chapter 8.

5.7 A Real-Life Analysis: Averting a Potentially Devastating Plant Consequence

Figure 5.5 is a typical schematic for a service water system. Service water by itself would most likely be a discarded system according to most RCM programs and RCM "consultants." The 80/20 truncated RCM rule, whereby 80 percent of the plant is omitted from consideration and only 20 percent is analyzed, and the other streamlined versions of RCM would probably never look at the components in this system. How wrong that would be. Any RCM program that would *not* consider analyzing this system would miss a major plant consequence. A very significant finding in a real-life RCM analysis was identified in this system, and that is precisely the reason I am using this real-life example that we will analyze together. I briefly reviewed this scenario in Chapter 1, but now let's take a closer look at it with a deeper understanding of what we have learned since Chapter 1 and apply our RCM analysis skills.

The service water system receives its normal service water supply from the local city water district. The system also has two service water pumps that can supply service water from an alternate storage tank if the city water supply should undergo repairs and be unavailable as a water source. The majority of components in this system operate to supply service water to lavatories, drinking fountains, and other relatively unimportant and noncritical functions, except for one very critical function.

The entire service water flow path extends to several additional drawings, so for clarity I have redrawn and simplified this flow path, which is shown in Figure 5.6. Let's commence our analysis with a rather innocuous check valve located in the

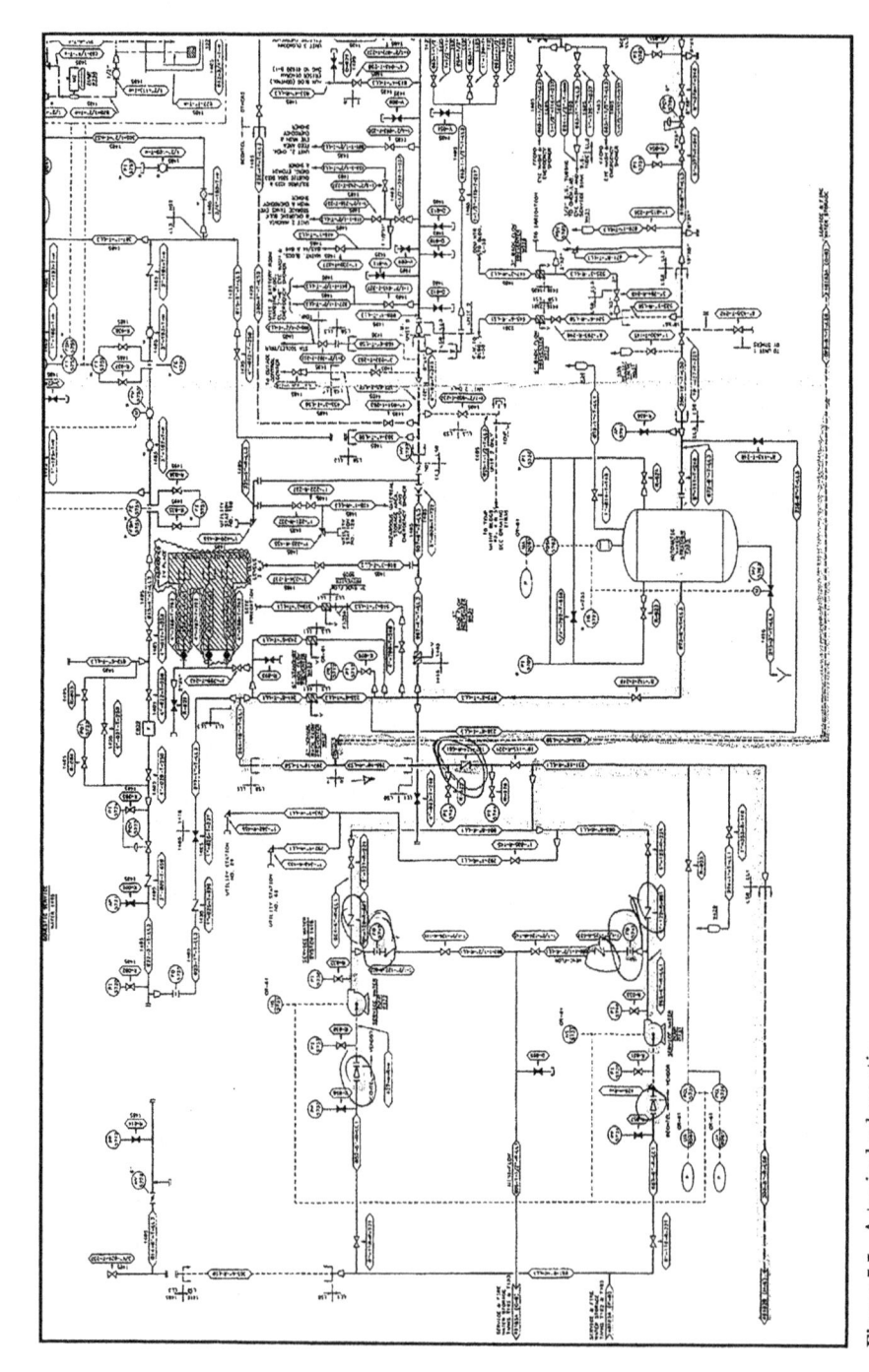

Figure 5.5 A typical schematic.

center of the schematic in Figure 5.5, which is depicted as check valve C in Figure 5.6. This is where we begin to enter our data on the COFA Worksheet for service water check valve C. Refer to Figure 4.2. We also need to refer to the COFA Logic Tree, the Potentially Critical Guideline, and the Economically Significant Guideline, as shown in Figures 5.2*a* and 5.2*b*. Each step of the COFA Worksheet is explained as follows. The first column on the COFA Worksheet identifies the component I.D. and description.

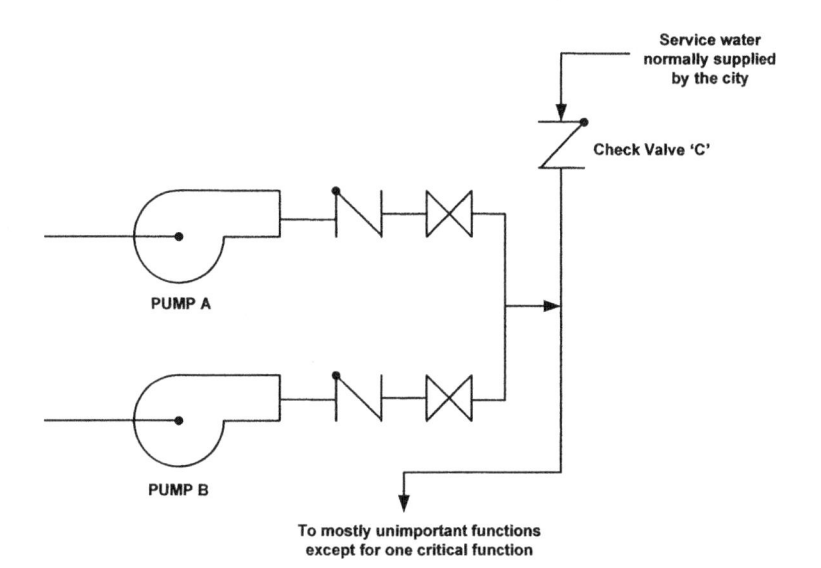

Plant Conditions:
o During normal plant operation service water is supplied by the city.
o Check valve 'C' fails in the OPEN position due to disc binding.

Analysis Considerations:
o The failure of check valve 'C' in the OPEN position is 'hidden' since there is no indication of failure. Also, there is no consequence of its failure in the OPEN position as long as service water is being supplied by the city.
o Even when pumps 'A' and 'B' are being periodically tested, a failed open check valve would still be hidden as long as service water was being supplied by the city.
o However, with an additional failure such as the rupture of the normal city water line to the plant caused inadvertently by a backhoe operator during construction, city maintenance, or repair, there will be a diversion of service water to a critical plant function.

Analysis Result:
The effect of check valve 'C' failing in the OPEN position during normal plant operation is considered to be "Potentially Critical" because its failure is "hidden" and a plant consequence will not occur <u>until</u> the second additional failure occurs. In essence, a remote city construction worker could cause the unplanned shutdown of two generating units.

Figure 5.6 Averting a potentially devastating consequence.

Component I.D. and description.
Valve I.D.: C
Service Water System
Check Valve

In column B of the COFA we describe all functions of the component.

Describe all functions of the component.
1. Provide a flow path from the city water supply to the service water system to provide service water to the lavatories and drinking fountains and to the bearings of the circulating water pumps.
2. Provide isolation from the service water system when the city water supply is unavailable.

In column C of the COFA we describe the ways that each function can fail.

Describe the ways that each function can fail.
1. Fails to provide a flow path from the city water supply to the service water system to provide service water to the lavatories and drinking fountains and to the bearings of all circulating water pumps.
2. Fails to provide isolation from the service water system when the city water supply is unavailable and the service water pumps are in operation.

In column D of the COFA we describe the dominant failure modes for each functional failure.

Describe the dominant component failure modes for each functional failure.
1. Check valve fails closed.
2. Check valve fails open.

In column E we determine whether the occurrence of the failure is evident. Refer to the COFA Logic Tree in Figure 5.2*a*.

Is the occurrence of the failure mode evident?
1. The occurrence of the check valve failing closed will be evident by low flow instrumentation in the control room.
2. The check valve failing open will not be evident. It will be hidden because as long as the city water supply is providing the flow of water, the check valve failing in the open position will not be evident. There is no indication of failure and there is no system operating consequence if the valve fails open. Even when pumps A and B are periodically run, the failed open check valve would still not be evident as long as service water pressure was available from the city water supply.

In column F we describe the system effect for each failure mode.

Describe the system effect for each failure mode.
1. If the check valve fails closed, it will be evident and the service water pumps will be available to supply the service water; therefore, there is no system effect.
2. If the check valve fails open, it is hidden but there is no system effect as long as service water continues to be supplied by the city water system.

In column G we describe the consequence of failure based on the asset criteria specified. Refer to the COFA logic and Potentially Critical Guideline in Figures 5.2a and 5.2b.

Describe the consequence of failure based on the asset reliability criteria specified (refer to the COFA Logic Tree).
1. Per the COFA Logic Tree, there is no safety, operability, or economic consequence of failure. Therefore, based on failure mode 1, this is a run-to-failure component.
2. The failure is hidden and per the COFA Logic Tree, we are directed to the Potentially Critical Guideline, which asks, "Can the component failure, in combination with an additional failure or initiating event, result in an unwanted consequence that has a direct adverse effect on one or more of the asset reliability criteria?"

Let's find out the answer to this question. With check valve C failed in the open position, we determined that there would be no system effect because the service water would always be supplied by either the city or the service water pumps via the alternate service water tank.

However, what would happen if there was an additional failure, such as a rupture of the city waterline? This could easily happen if a worker on a construction crew or a city employee working for the municipality itself were operating a backhoe and inadvertently ruptured the city water supply line to the plant. Looking at the simplified schematic in Figure 5.6, the two service water pumps would start, but there would be a diversion of service water flow from the pumps through the failed open check valve. This in turn would result in the loss of bearing seal water to all of the circulating water pumps of both units. Therefore, the consequence of that additional second failure in combination with the failed open check valve would be a dual unit shutdown! The answer to the question in the Potentially Critical Guideline is emphatically *yes!* The answer to the question in column G is a dual unit shutdown of two generating units simultaneously!

This true scenario shows how a rather innocuous component, whose failure was hidden, making it even more innocuous, has the potential to cause a serious consequence. This also illustrates how the same component can have one failure mode that would be classified as run-to-failure, but its other failure mode is potentially critical.

In column H we define the component classification.

Describe the component classification.

The component is classified as potentially critical due to a possible simultaneous dual unit shutdown.

Obviously, this was a disaster waiting to happen. As long as an incompetent backhoe operator never ruptured the city waterline while check valve C was failed open, this devastating event would not happen and no one would ever recognize this vulnerability. But just imagine if the city waterline was inadvertently ruptured while check valve C was failed open in its sleeper cell mode. As I mentioned several times in previous chapters, it is not the obvious that causes the greatest disasters; it is the *non*obvious! And it is usually only a matter of time before Murphy's Law takes effect.

If this calamity had occurred, it would have attracted the interest and inquisitiveness of the CEO, the maintenance supervisor, and everyone under him or her, who would have to explain the occurrence. A root-cause evaluation, probably no less than 6 inches thick, entailing hundreds of labor-hours to prepare, would appear most expeditiously.

This true-life example also highlights the importance of the concept of potentially critical components. Most RCM programs would not have identified this vulnerability. Finding this vulnerability and establishing a preventive maintenance strategy or possibly implementing a design change *before* the consequence occurs is what *RCM Implementation Made Simple* is all about!

5.8 Why Streamlined RCM Methods Are Not Recommended

The case of the service water check valve analyzed here is just a typical example. If you are going to expend effort to enhance your preventive maintenance program, why settle for only a marginal improvement when basically the same effort could reap a maximum enhancement? Preventing a dual unit shutdown, in itself, would be reason enough not to pursue streamlined versions that would have virtually no chance of finding that vulnerability. There are many more of these situations just waiting to happen. Some of the common adaptations of streamlined RCM are Total Productive Maintenance (TPM), Reliability-Based Maintenance (RBM), Probabilistic Safety Analysis (PSA) Based Maintenance, and the 80/20 Rule. These are acceptable if you want to accept major risks and do not want to ferret out the true vulnerabilities of your facility. Let's briefly look at the general characteristics of each one; the shortcomings will become obvious.

5.8.1 Total productive maintenance (TPM)

This streamlined version includes ownership teams comprising representatives from operations, maintenance, and engineering. These teams determine what equipment gets worked on and when it gets scheduled. Comprehensive adherence to any formal analysis logic is missing. This is a form of best-guess maintenance. Granted, some team members may have significant experience with some of the equipment, but that is as thorough and accurate as it gets.

5.8.2 Reliability-based maintenance (RBM)

This version starts with an assessment of the current PM program, and some maintenance visualization is made of what the PM program should look like in the future. Then RCM is employed on a hit-or-miss basis on the components that were part of the visualization.

5.8.3 Probabilistic safety analysis (PSA) based maintenance

This adaptation employs probabilistic methods usually associated with a specific safety significance. Typical operability consequences and personnel safety issues are not normally included in the PSA model. That is the extent of the rigor. As an example, PSA is used primarily to determine the probability of a core melting down, which is an extremely important safety consideration in a nuclear environment. But the robustness of PSA to the remaining components and systems not directly associated with a core meltdown is absent.

5.8.4 80/20 rule

This rule is definitely high risk. Vying for this option of streamlining your program is not even RCM in its most remote sense. This is where 80 percent of the plant is ignored and only 20 percent of the plant is analyzed because the proponents of this version believe only 20 percent of the plant is important enough to be evaluated. I liken this to buying car insurance that insures you only while you are driving 65 mph on a freeway or while you are driving in heavy traffic, when an accident is most likely to occur. You would not be insured driving on country roads, or driving slowly through your neighborhood to and from work, or driving on any nonbusy roads because it is assumed you would not have an accident under these conditions. You would assume the risk of having no insurance coverage during these times. Does that sound comforting?

Many people think that as long as they use the proper RCM logic somewhere in their analysis, they can streamline even the classical version of RCM, whether that classical approach started at the system level or at the component level. That is *not* RCM, rather, it is a pick-and-choose approach involving only the systems or components they want to analyze.

I believe that the streamlined versions of RCM only reinforce what is already known. Every facility has its "known" problem systems and problem components for which failures can result in undesirable plant consequences. The identification of this equipment population does not require much insight. Experience has already seen to that.

It is also my personal belief that in addition to being attuned to the known problem systems and problem components, *plant reliability and safety are directly related to the vulnerabilities that have* NOT *yet been identified because the failure consequences surrounding those vulnerabilities have not yet occurred.* Streamlined versions of the RCM process will most likely result in those plant vulnerabilities *remaining* unidentified until . . .

RCM Implementation Made Simple is all about finding those vulnerabilities *before* they occur and result in an unwanted consequence of failure.

5.9 Chapter Summary

Once again, we covered a lot in this chapter, but in a very logical, straightforward manner, so let's review the highlights. Refer to Figures 5.7*a* and 5.7*b*.

144 Chapter Five

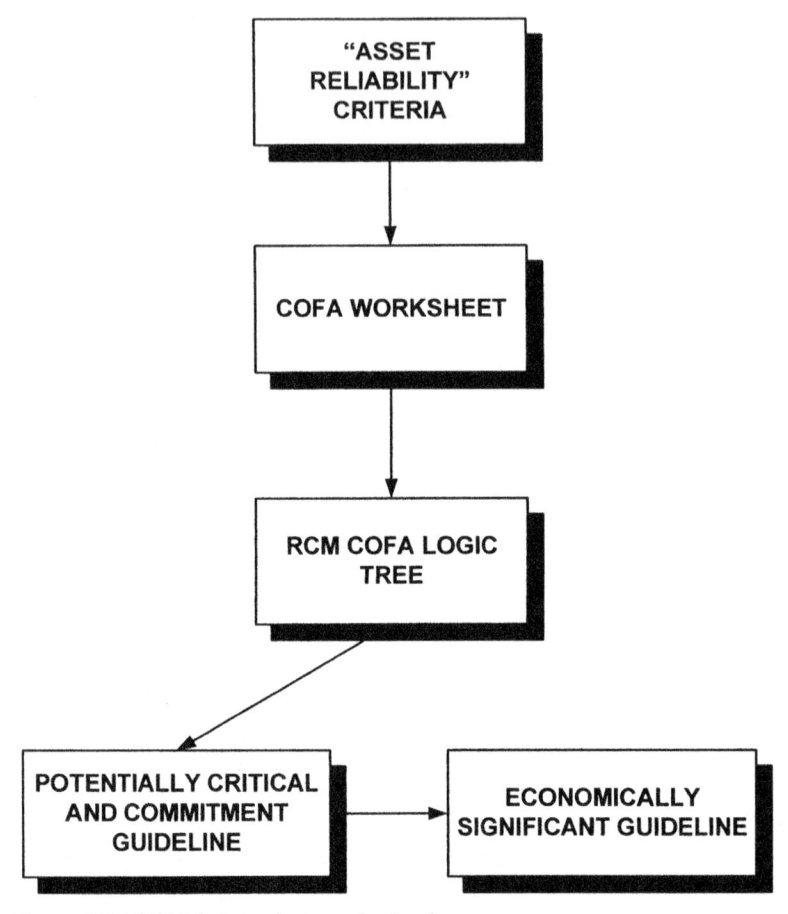

Figure 5.7(a) RCM decision logic made simple.

- RCM is the *driver* for a corporate reliability strategy.
- The first step in implementing your RCM program is to define precisely, accurately, and exactly the criteria your senior management wishes to preserve. It is strongly advised that you submit a formal internal memorandum for signatures from all applicable individuals in authority for agreement on what you choose.
- Figure 5.1 lists typical asset reliability criteria. *Personnel, public, and plant safety* criteria are *mandatory*. The remaining criteria are operability criteria.

Figure 5.7(b) RCM logic sequence for identifying the consequence of failure.

- It is perfectly acceptable to place qualifying conditions on the criteria you select, such as a production or an assembly-line interruption with a duration of no more than 15 minutes, or 30 minutes, or 1 hour, or a power reduction of no more than 1 percent, 5 percent, or 10 percent. Any qualifying conditions you choose should also have the buy-in of senior management.
- Your PM program will consist of components classified as either critical safety or operability, potentially critical safety or operability, commitment, or economic.
- The RCM COFA Logic Tree is truly *RCM Implementation Made Simple.* A complex logic process has been simplified to its basic elements while enhancing its thoroughness, accuracy, and clarity.
- The Potentially Critical Guideline and the Economically Significant Guideline are integrated with the COFA Logic Tree to maintain the simplicity for identifying hidden failure consequences and purely economic consequences.
- The COFA Worksheet begins by describing *all* functions of the component.
- There is a functional failure corresponding to each function of the component.
- There is a failure mode corresponding to each functional failure. A component will have an associated failure mode for each different functional failure.
- The failure modes are those that are dominant, or most likely to occur. They do not include implausible or unrealistic failure modes. Failure modes are the types of failure or the ways a component can fail. Failure modes would include, for example, "valve fails closed," "valve fails open," "switch fails to actuate," "switch actuates prematurely," or "pump fails to start."
- There is not a system effect for every failure mode.
- There is not an unwanted consequence for every failure mode.
- A component with a failure mode that results in an unwanted consequence is classified as either critical, potentially critical, commitment, or economic.

- RCM translates the design functions of a component into the operating objectives of the facility via a reliability program.
- *RCM Implementation Made Simple* is in total concert with, and in many aspects is more advanced than, the SAE Standard for RCM as delineated in SAE Document JA1011.
- Streamlined versions of RCM fall far short of being sufficiently robust and comprehensive to identify the real vulnerabilities of your plant or facility. Plant reliability and safety are related to the vulnerabilities that have *not* yet been identified because the failure consequences surrounding those vulnerabilities have not yet occurred.

5.10 RCM Made "Difficult"

Now that's an odd subsection, you might be thinking to yourself. It is. Nonetheless, in Chapter 3 I mentioned that I would include the old-fashioned way of developing an RCM program for those readers who would prefer using the FMEA format to identify functions at the system level and establish boundaries and interfaces that are a required part of the analysis if you choose to identify system functions rather than component functions using the COFA format. Either way is correct. However, it is much more difficult and requires much more RCM expertise to define system boundaries, interfaces, and functions at the system and subsystem level. If you have that expertise, the more difficult method of RCM may be appropriate for you.

The differences between the two methods are analogous to the following scenarios. Some individuals prefer to use the old-fashioned slide rule rather than an electronic calculator. There is nothing wrong with using a slide rule; after all, it was in vogue for many, many years. Even though you have to manually decide where the decimal point goes and you have to keep flipping it over to arrive at an integer, it is an acceptable method to use. Similarly, some people feel more comfortable using an old-fashioned plug-in phone. They prefer knowing that the cord is attached. Others, however, have progressed to cordless cell phones. Either phone works just fine, though.

Regardless of whether you choose the old-fashioned method of classical RCM or the simple method of classical RCM, there are no shortcuts allowed. The streamlined versions whereby you pick and choose specific systems or components to analyze are not sanctioned in either case.

For those readers who want to stick with the FMEA format, developing your classical RCM program must include the following steps:

5.10.1 Determine system boundaries

The plant must be divided into discrete but arbitrary systems, which includes marking the plant P&IDs accordingly. The arbitrary boundary envelops the entire system and is shown on the P&IDs. The boundary of a system must include everything necessary for the system to accomplish its function.

The system boundary points are noted on the respective P&IDs and are then identified on the interface sheets. Boundary demarcations are drawn such that the controlled components and their associated controllers and instrumentation are within the system boundary. Boundaries are usually drawn at a valve, with the valve included as part of the system being analyzed, if its function is for isolation of the system. There are nominally several dozen arbitrary boundary points that must be identified for each system to enclose that discrete system. Sometimes boundaries are extended beyond that shown on a given drawing to include components that are integral to the system logic. All components within the system boundary are included as part of the database for that specific system. *A component can reside in only one system.*

5.10.2 Determine subsystem boundaries

Plant systems are frequently composed of a large number of components that serve a variety of functions to support the total operation of the system. Partitioning into subsystems refers to establishing groupings of components that are related to performing a particular function within the system. All instruments and components that are necessary for the subsystem to perform its function are included within the subsystem boundary. The primary reason for partitioning into subsystems is so you can analyze smaller slices of a larger system. There are numerous subsystem arbitrary boundary points that must be identified for each subsystem to enclose that discrete subsystem within the larger overall system. *A component can reside in only one subsystem.*

5.10.3 Determine interfaces

Establishing boundaries also includes identifying significant mechanical, electrical, and pneumatic inputs and/or control signal interfaces. These inputs are necessary for the subsystem components to function properly. All boundary points are specified as either *in-system* boundary interfaces or *out-system* boundary interfaces. If the boundary point component is supplying a function from another subsystem to the subsystem being analyzed, it is an *in-system* boundary interface. If the boundary point component provides a function from the subsystem being analyzed to another subsystem, it is an *out-system* boundary interface. Out-system boundary interfaces belong to the system being analyzed. In-system interfaces do not belong to the subsystem being analyzed. They belong to the subsystem governing their function. The boundary interface components are identified to delineate the exact demarcation of each system and subsystem so that when analyzing other systems, no component will be missed in the analysis.

5.10.4 Determine functions

When systems are partitioned into smaller subsystems, the system functions are specified at the subsystem level. Determining subsystem functions is an important step in the RCM analysis, since preserving these functions is the objective of the preventive maintenance program. Function definitions describe what the system or subsystem must accomplish. Criteria for defining functions can be ascertained by some of the following:

- Subsystem in-interfaces that must be supported
- Subsystem out-interfaces that are provided to another system
- Internal interfaces that the subsystem provides as input to another subsystem within the same system

An example of an in-interface that must be supported is as follows:

> When analyzing the circ pump subsystem of the circ water system, an in-interface is service water supplied by the service water system. This in turn requires that a function be defined when analyzing the service water system to "provide service water to the circ water pumps."

Typical examples of subsystem functions could include the following:

- "Provide sufficient flow to the main condenser."
- "Provide screen wash for cleaning the traveling bars and screens."

An example of an internal interface provided as input to another subsystem within the same system is as follows:

> In the circ pump subsystem, an internal interface is the interlock signals provided from the condenser and discharge subsystem of the same circ water system. These interfaces to the circ pump subsystem are functions of the condenser and discharge subsystem.

5.10.5 Determine the functional failures

Determining the functional failures is the same process used in the simplified classical RCM process. It is basically the opposite of the function. For example, if the function reads "provide sufficient flow to the main condenser," the functional failure will be written as "fails to provide sufficient flow to the main condenser."

5.10.6 Determine which equipment is responsible for the functional failures

This is the step in the process where you need to determine precisely what component I.D. is responsible for the functional failure. You need to ascertain which component it is whose failure results in a failure of the function.

The remaining steps of determining dominant failure modes, system effects, and plant consequences are similar to the process described earlier in this chapter.

This gives you a very clear indication of why I wrote this book—and why *RCM Implementation Made Simple* is so much more robust and easier to understand and implement.

In the next chapter we will look at the PM task selection process.

Chapter

6

The PM Task Selection Process

This chapter explains the strategies for phase 2 of an RCM program. This is where preventive maintenance tasks are established to address the causes of failure for the equipment population identified in phase 1 (of the three phases of an RCM program). The PM tasks are relevant to components that we learned how to recognize via the COFA, the Potentially Critical Guideline, and the Economically Significant Guideline. They are the components classified as critical, potentially critical, commitment, or economic.

I have intentionally separated this element of the process from the element for defining the component classification for several reasons, but mainly because it makes the entire analysis simpler to understand. Secondly, as I mentioned earlier, these two elements require different mind-sets and different expertise to obtain the optimum accuracy and maximum thoroughness of the process. Nevertheless, if you prefer to combine everything and complete the component classification, the PM task, and the periodicity elements at the same time, that is a totally acceptable approach. I have found, however, that doing so disrupts the momentum of the thought process and interferes with the efficiency of completing the analysis. It is not that different from building a house. There are certain elements of the building process that are more efficient and provide a better outcome when they are sequenced.

6.1 Understanding Preventive Maintenance Task Terminology

To begin, a basic understanding of the definitions of preventive maintenance tasks is necessary to avoid confusion about the many different terms used in the industry. Terminology associated with preventive maintenance tasks includes the following: *time-directed, condition-directed, condition-based, proactive, reactive, predictive, failure-finding, in situ, on-condition,* and *surveillances.* These terms mean different things to different people, and I am not advocating that any specific definition be used. However, I have chosen to use the fundamental RCM terms for the different categories of preventive maintenance, which is the convention used in this chapter and throughout this book to describe preventive maintenance activities. There are also a myriad of different types of PM activities, such as the following: *overhauls, inspections, performance tests, bench tests, oil sampling, thermography, vibration analysis, motor current signature analysis (MCSA), eddy current testing, hi-pot testing, calibrations, monitoring, replacements, disassemblies, cleaning, nondestructive testing (NDT),* and *acoustics.*

People who are familiar with preventive maintenance might use these terms differently or interchangeably to mean the same thing. Therefore, to avoid confusion, I use these terms for the three basic categories of preventive maintenance tasks: condition-directed, time-directed, and failure-finding.

6.2 Condition-Directed, Time-Directed, and Failure-Finding Tasks

Throughout this chapter, preventive maintenance tasks consist of three general categories: condition-directed, time-directed, and failure-finding. The different *types* of PM activities fall into one of these three categories. Condition-directed and time-directed preventive maintenance is specifically intended to *prevent* failures at the *component level.* These tasks address the different causes of failure in order to prevent equipment failures from occurring. Failure-finding preventive maintenance tasks do not prevent failures at the component level. Failure finding is a strategy to ascertain, at a periodic interval, whether or not a specific component or system has *already* failed so that the failed component or system can be detected *before* it results in a plant consequence upon the occurrence of an additional failure or initiating event. Therefore, failure-finding tasks can be viewed as a preventive maintenance strategy to prevent failure consequences at the *plant level*. Failure-finding preventive maintenance is applicable to hidden failures and is also performed on safety systems and components that are not normally operating (they operate on demand). Refer to Figure 6.1.

Time-directed tasks normally include replacements, overhauls, and restoration of components at given periodicities. For the most part they are intrusive and require disassembly or removal. Condition-directed tasks normally include tasks that measure, monitor, or analyze the condition of a component to determine whether it is operating acceptably or is about to fail. Predictive maintenance (PdM) tasks such as vibration monitoring, oil analysis, thermography, and MCSA are all types of tasks that fall into the condition-directed category. Of course, condition-directed and

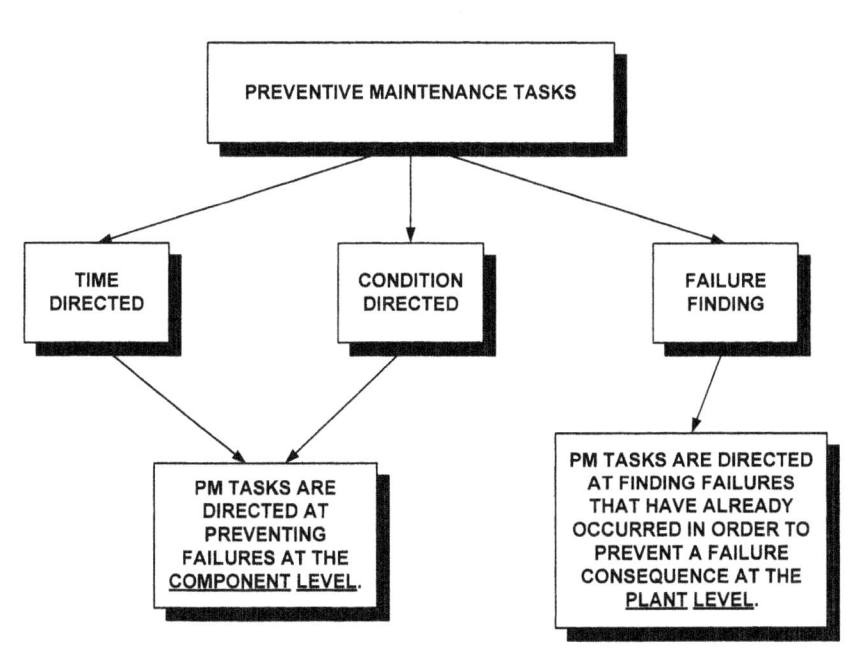

Figure 6.1 Preventive maintenance tasks.

failure-finding tasks still must be scheduled at some given periodicity, but that does not qualify them as time-directed.

There is a prevalent but misguided belief on the part of many engineers and senior-level managers regarding condition-directed tasks. They still cling to the notion that even though a component has a condition-directed task specified but not a time-directed overhaul or replacement, it is a run-to-failure component. This is completely incorrect. Condition-directed maintenance is *not* run-to-failure. Condition-directed maintenance, according to its most basic definition, means . . .

> *Don't overhaul or replace it until its condition indicates the need for overhaul or replacement.* Predictive maintenance techniques are used to determine the condition of the equipment so that required overhaul or replacement can be scheduled to preclude the occurrence of a functional failure.

Quite often the term *proactive preventive maintenance* is used to connote taking action before failures occur, as opposed to *reactive maintenance,* which implies that no action is taken until after a failure occurs. Many people believe that proactive maintenance is some new form of maintenance that just came on the scene in the past few years. Proactive preventive maintenance is really synonymous with *predictive* maintenance, which is a subset of condition-based maintenance. Therefore, proactive preventive maintenance is not really new. Any newness comes from the newer PdM techniques and the relatively new emphasis placed on PdM. The RCM philosophy has always been proactive, as advocated in Nowlan and Heap's work in 1978.

Another common misconception is that all preventive maintenance is performed by the maintenance department. In fact, preventive maintenance is also performed by operations, engineering, chemistry, and other departments. Refer to Figure 6.2. When we get to the PM Task Worksheet, all applicable and effective tasks will be described, including those performed by operations, engineering, and others, as long as those activities are formally documented and proceduralized. Note that Figure 6.2 uses the expressions "Operations PM Tasks" and "Engineering PM Tasks." Some of the activities these organizations perform are indeed PM tasks. Maintenance is not the only department that performs preventive maintenance. If operations is already

The PM Task Selection Process   157

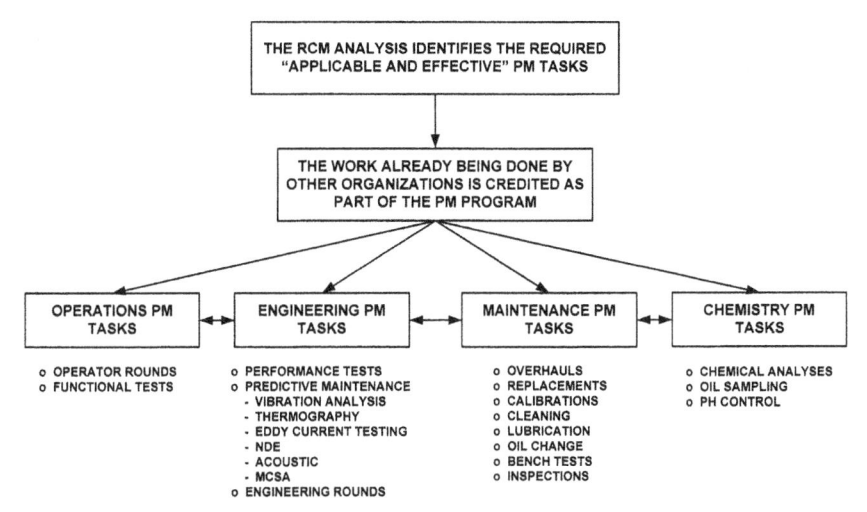

Figure 6.2 Departmental integration for preventive maintenance optimization.

performing routine inspections during their daily rounds, why duplicate that effort with another maintenance task to accomplish the same thing?

Some industries use the term *RM*, or *repetitive maintenance*, to identify all repetitively scheduled tasks, of which PMs are just one type. If a regulatory required task is repetitively scheduled, it may be called an SV for a surveillance or a TS for a technical specification. These are all subsets of RM. Since each industry has its own idiosyncratic ways of defining repetitive tasks and the associated terms, I have kept *RCM Implementation Made Simple* as just that—simple. Therefore, I refer to a PM for any repetitively scheduled task and a CM for a corrective maintenance activity.

6.3 The PM Task Worksheet

The PM tasks are entered on the PM Task Worksheet, which is one of the implementation tools shown in Figure 4.3. The data for columns A through C of the PM Task Worksheet comes directly from the COFA Worksheet. The component I.D. and description, the consequences of failure, and *each* dominant failure mode are entered.

In column D of the PM Task Worksheet, the credible cause of failure must be identified for *each* component failure mode that resulted in an unwanted consequence for any of the asset reliability criteria regardless of whether it was critical, potentially critical, commitment, or economic. Each failure mode resulting in one of these classifications must have a preventive maintenance task specified to prevent its cause of failure. Invariably, there will be *several* causes of failure for each failure mode.

It is important that each failure mode be addressed because in one instance a component may fail in a certain way and the PM task specified to prevent that cause of failure may be different from a PM task to prevent a different cause of failure for a different failure mode if the component failed in a different manner. More simply stated, a component can fail open or closed, or operate prematurely, or fail to start, or stick, or bind, and there will be different causes for these different failure modes, which in turn require different PM tasks to address these different causes.

The cause of failure in most instances involves a straightforward process. However, it is recommended that the failure causes be determined by knowledgeable individuals with a thorough understanding of the equipment. These individuals should be representatives from within the maintenance and engineering organizations who are familiar with the specific equipment. Operations personnel may also have an input; however, maintenance and engineering types are usually more intimately familiar with internal failure mechanisms of the equipment than the operators.

For example, the causes for a motor-operated valve failing open (or failing closed) could be a failed motor, a failed torque switch, a failed limit switch, a bearing failure, an internal solenoid failure, and so on. The causes of an electric pump failure could be failed bearings, failed stator windings, a shaft failure, excessive impeller wear, overheat switch failure, internal electrical circuitry control failure, and so on. In many instances, a single task will address several failure causes.

6.4 The PM Task Selection Logic Tree

Before we enter the "applicable and effective" PM tasks in column E to address and prevent the causes of failure for each failure mode, an understanding of the task selection process is needed; I will explain this process in detail. Refer to the PM Task Selection Logic Tree in Figure 6.3. The first option for a PM task is a condition-directed task. This is a nonintrusive task such as implementing a predictive maintenance technology or performing an external inspection or a performance test. A nonintrusive task is always preferable. The second choice is a time-directed task, which is usually an intrusive task such as an overhaul, a

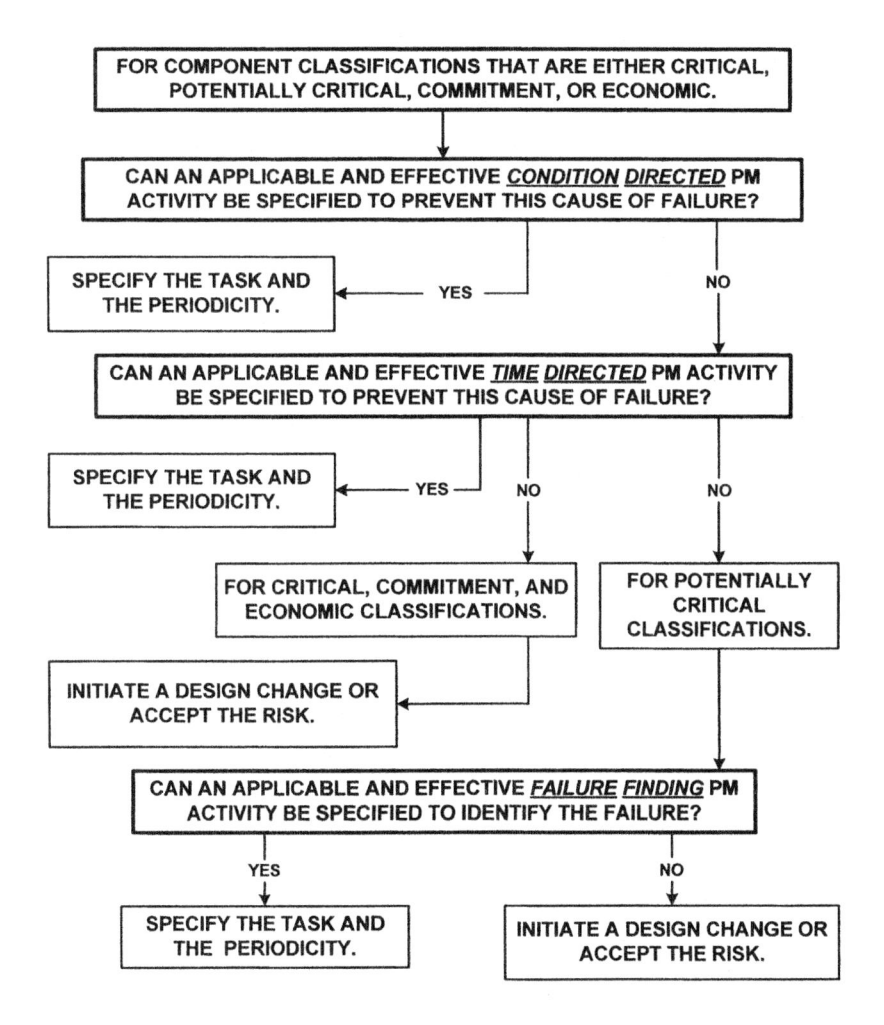

```
┌─────────────────────────────────────────────────────────────┐
│ FOR COMPONENT CLASSIFICATIONS THAT ARE EITHER CRITICAL,     │
│ POTENTIALLY CRITICAL, COMMITMENT, OR ECONOMIC.              │
└─────────────────────────────────────────────────────────────┘
                              ↓
┌─────────────────────────────────────────────────────────────┐
│ CAN AN APPLICABLE AND EFFECTIVE CONDITION DIRECTED PM       │
│ ACTIVITY BE SPECIFIED TO PREVENT THIS CAUSE OF FAILURE?     │
└─────────────────────────────────────────────────────────────┘
           ↓ YES                              ↓ NO
┌──────────────────────┐
│ SPECIFY THE TASK AND │
│   THE PERIODICITY.   │
└──────────────────────┘
┌─────────────────────────────────────────────────────────────┐
│ CAN AN APPLICABLE AND EFFECTIVE TIME DIRECTED PM ACTIVITY   │
│ BE SPECIFIED TO PREVENT THIS CAUSE OF FAILURE?              │
└─────────────────────────────────────────────────────────────┘
     ↓ YES              ↓ NO                  ↓ NO
┌──────────────────────┐
│ SPECIFY THE TASK AND │
│   THE PERIODICITY.   │
└──────────────────────┘
         ┌────────────────────────────┐  ┌──────────────────────┐
         │ FOR CRITICAL, COMMITMENT,  │  │  FOR POTENTIALLY     │
         │ AND ECONOMIC CLASSIFICATIONS.│ │      CRITICAL        │
         │                            │  │  CLASSIFICATIONS.    │
         └────────────────────────────┘  └──────────────────────┘
┌──────────────────────────────┐
│ INITIATE A DESIGN CHANGE OR  │
│       ACCEPT THE RISK.       │
└──────────────────────────────┘
┌─────────────────────────────────────────────────────────────┐
│ CAN AN APPLICABLE AND EFFECTIVE FAILURE FINDING PM          │
│ ACTIVITY BE SPECIFIED TO IDENTIFY THE FAILURE?              │
└─────────────────────────────────────────────────────────────┘
           ↓ YES                              ↓ NO
┌──────────────────────┐         ┌──────────────────────────────┐
│ SPECIFY THE TASK AND │         │ INITIATE A DESIGN CHANGE OR  │
│   THE PERIODICITY.   │         │       ACCEPT THE RISK.       │
└──────────────────────┘         └──────────────────────────────┘
```

NOTE: A FAILURE FINDING TASK IS NOT APPLICABLE FOR CRITICAL, COMMITMENT, OR ECONOMIC COMPONENTS BECAUSE THE CONSEQUENCE HAS ALREADY OCCURRED WHEN THE COMPONENT FAILED.

Figure 6.3 PM Task Selection Logic Tree.

replacement, an internal inspection, or the like. Depending on whether or not a condition-directed or a time-directed task is applicable and effective, a design change may be required.

Note also that the PM task must be applicable and effective. In order to be applicable and effective, the proposed preventive task must be such that it can be appropriately applied to the component. It must be appropriate for addressing the failure cause. The preventive task must offer some degree of assurance that it will prevent or at least minimize the exposure to a failure of the component. The selected task, based on a principle of prudent judgment by knowledgeable individuals, should have a relative degree of pertinence and likelihood that it will prevent the occurrence of the failure mechanism. You should ask, "Does this task really make sense, or is it being identified only for the sake of 'doing something' without regard for the criteria?"—in which case, it would not be applicable and effective.

I have seen many instances where equipment failures occurred and totally ineffective PMs were created only for the sake of convincing management that something was being done about the failures. Such PMs are not applicable and effective. In those cases, only a design change or an upgraded component is the answer. If that is the conclusion you come to, then don't shy away from it.

Prudent judgment must prevail when identifying credible failure causes and applicable and effective PM tasks. In some instances it is appropriate to accept the fact that an applicable and effective *preventive* task cannot be identified. In such a case, either a failure-finding task or a design change may be warranted.

A failure-finding task is applicable only to potentially critical components because the unwanted consequence of failure for critical, commitment, or economic components has already occurred when the component failed.

In the case of a hidden failure, if a preventive task cannot adequately ascertain a pending failure or prevent its occurrence, a failure-finding task must be specified. Failure-finding tasks ascertain only that a hidden failure has already occurred, but having that knowledge allows you to prevent an unwanted plant consequence *before* an additional failure or initiating event should occur.

6.5 Why a Condition-Directed Task Is Preferred

Why is the condition-directed task preferred? Figure 6.4 is a typical depiction of the "bathtub" curve, which is credited to the pioneers of RCM, Nowlan and Heap. There are several interesting attributes to note when looking at the bathtub curve. The most important one is that whenever a component is removed from service for an overhaul or replacement, its time is zeroed out but the failure probability increases dramatically for two reasons. One is premature failures, and the other is infant mortality. Therefore, it is more prudent and efficient to allow a component to operate until some predictive task or set of tasks shows that the component has impending failure mechanisms and is in need of overhaul or replacement.

Premature failures are very common. Quite often a piece of equipment that is operating satisfactorily will be removed from service to satisfy a vendor's recommendation for replacement at some given time interval. Once the equipment is replaced, there are a host of reasons for the newly overhauled unit to fail prematurely. There could be problems during its reassembly, especially if it is a rather complex component. There could be problems with

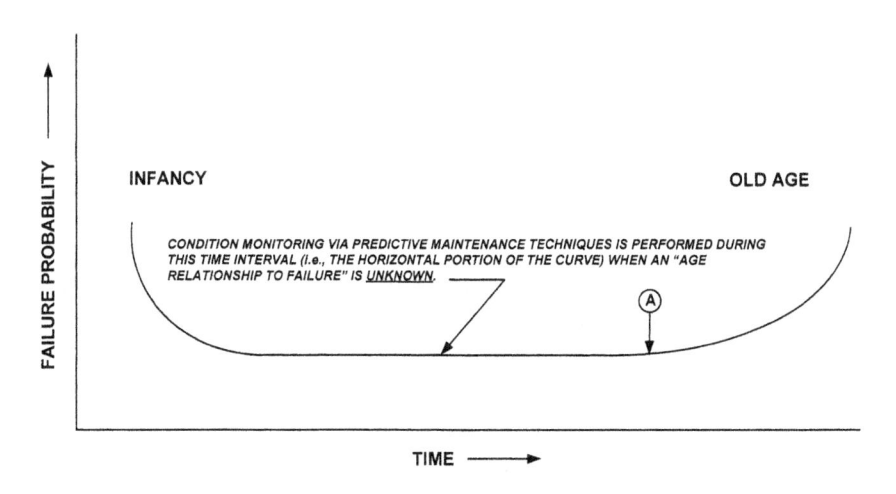

NOTE: EVEN WITH A _KNOWN_ "AGE RELATIONSHIP TO FAILURE" AS IDENTIFIED BY POINT 'A', PREDICTIVE TECHNIQUES (PdM) ARE STILL PERFORMED TO IDENTIFY ANY PREMATURE FAILURE MODES.

Figure 6.4 The "bathtub" curve.

the replacement parts. There could be problems encountered when reinstalling the component itself.

Another source of infant mortality occurs with new equipment during its "burn-in" period. Many components require this burn-in period to allow clearances, wear rings, seals, and other piece parts to find their relaxed operating parameters.

Furthermore, if you replace a component before it becomes necessary, you will lose all of the remaining time along the horizontal portion of the bathtub curve until point A in Figure 6.4 is finally reached. For example, based on a vendor's recommendation, a component may be overhauled every 5 years, whether it needs it or not, when it could have lasted for 10 years or more. Not only is that a waste of resources, but it diminishes the available resources for other critical work.

There are some components that were designed to last the life of the plant, whether that is 20, 30, or 40-plus years. Everything else will eventually fail prior to that time, so why remove it until its condition indicates the need for replacement?

Another interesting attribute is that Nowlan and Heap determined that only approximately 11 percent of all components universally exhibit a given wear-out point, shown as point A in Figure 6.4. This means that approximately 89 percent of all components fail randomly, and that is why a time-directed overhaul without regard to the condition of the component is not a good practice. There are, however, many components (those in the 11 percent category) that exhibit an age relationship to failure, and a time-directed overhaul or replacement is the recommended task.

Oftentimes both a condition-directed task and a time-directed task are specified for the component for different causes of failure. For example, a condition-directed task may be specified to periodically monitor the component for vibration to ensure that an incipient bearing failure is not about to occur, while a time-directed overhaul may also be specified for the component if an age relationship to failure is known.

6.6 Determining the PM Task Frequency and Interval

Now we come to another milestone in the PM task selection process. How do you define the frequency and interval for the tasks?

The frequency and interval (i.e., the periodicity) for the tasks selected are entered in column F of the PM Task Worksheet. The frequency is usually expressed as either hours, days, quarters, months, annually, or years. The interval is expressed numerically. For example, a periodicity of once every four months would be M4, a periodicity of every four years would be A4 or Y4, depending on whether you choose to use an annual or a yearly expression. Determining the periodicity can be extremely complicated or it can be quite simple and based on common sense. I prefer the latter approach. But first let's explore the more complicated version so you understand why that is not the preferred method *except* for calculating the probability of failure, or the mean time between failures (MTBF).

Determining the probability of failure, or the MTBF, is an acceptable practice, since it is a calculation for estimating the operating life of a component. Similarly, any calculations—such as calculating the life of bearings, motor windings, pump wear, and so on—are all applicable for estimating the age relationship to failure for a component. Some intelligence, however, is still required when using these numbers. This becomes evident in the paragraph that follows.

The complicated method may involve other mathematical calculations, formulas, or statistical methods that are not easily developed. However, this is only a minor reason for not pursuing that route. More important, as you can see in Figure 6.5, PM task frequency and interval considerations are more of an art than a science. Just by reviewing these considerations, it is quite evident that some "exact" mathematical model could determine that 2.73 years is the optimum time for scheduling a periodic performance test of a specific component, for example. What if that component requires a planned outage to gain access to it, and the outage is scheduled every three years? Will the outage have to be scheduled sooner? What if "other" activities are performed on that component every two years? Will this require that 0.73 years after the component was worked on for some other PM activity, it must be tested again? This is not the way the real world of maintenance works. You should be as efficient as possible and schedule work on the component as efficiently as possible. Therefore, prudent engineering judgment would be more practical by considering all of the conditional parameters

- FAILURE HISTORY
- CM HISTORY
- VENDOR RECOMMENDATIONS
- INDUSTRY HISTORY
- REGULATORY REQUIREMENTS
- DESIGN AND OPERATING CONSIDERATIONS
- OTHER TASKS SCHEDULED ON THE SAME COMPONENT
- PLANNED OUTAGES
- ABILITY TO GAIN ACCESS TO THE COMPONENT
- OPERATOR CAPABILITIES
- PdM MONITORING ACTIVITIES
- ENVIRONMENT

→ OPTIMUM PM TASK FREQUENCY AND INTERVAL

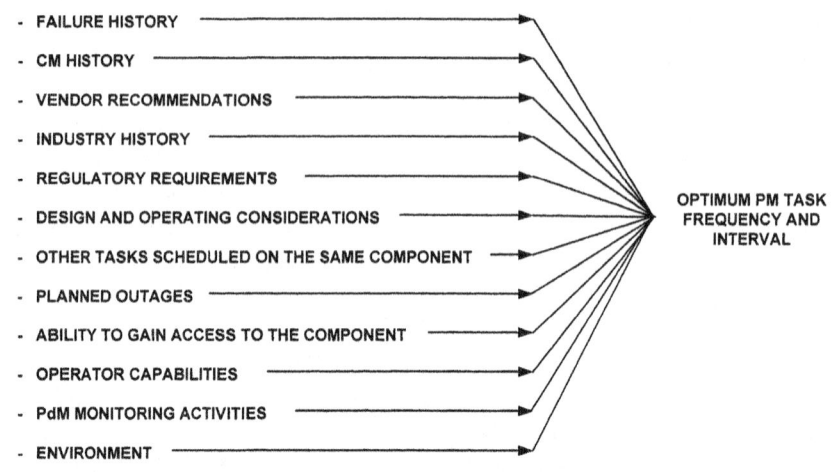

Figure 6.5 PM task frequency and interval considerations.

and determining an *optimum* periodicity. These are a few of the reasons that prudent judgment on the part of individuals knowledgeable about the equipment is the recommended method for establishing task periodicities. Of course, if you have good failure history data, that data should always be a primary consideration for establishing periodicities.

In the real world, a plant is designed to operate according to certain efficiency standards, with operational run times that can produce manufactured output such as a certain number of computer chips per day, process output such as a certain number of barrels of oil per day from a refinery, or generate output such as a certain number of cubic feet of natural gas per day or a certain amount of megawatts of electricity per hour. A reliability program is designed to ensure that the facility operates reliably to meet those goals. The plant was not designed to operate within the parameters of a reliability program. Therefore, the RCM program must conform to certain constraints such as when the equipment is available to be worked on.

Most times, operating equipment cannot be worked on for maintenance unless the component or system can be isolated and cleared for being worked on. This is not always a simple objective to meet. Therefore, a facility will usually have a long-range schedule of approximately one to two years, or even longer, whereby planned maintenance is scheduled during a complete overhaul of the entire plant or during an overhaul of sections of the plant at a given time. Usually, a scheduled rotating basis is established for working on the equipment in the various systems of a facility. These are the planned "windows of opportunity" to perform most maintenance work—especially intrusive maintenance activities. Obviously, any unplanned equipment failure always has the potential to modify that schedule immediately. It is important to consider such things as when your plant outages are scheduled and when the next production line outage is scheduled when selecting the appropriate time for working on components that cannot be isolated for access at other times.

When it has been determined that the periodicity at which the equipment needs to be worked on and the scheduled outages do not coincide, prudent judgment by individuals knowledgeable about the equipment and the plant is once again the recommended method for either adjusting task periodicities or changing plant outage intervals.

This is why selecting the appropriate PM task and periodicity is more art than science. In Chapter 8, we will find out how to make prudent adjustments to the task intervals that we initially established when we explore how to develop an RCM "living program."

6.6.1 The optimum time to establish a reliability program

The optimum time to implement an RCM program is during the design stage of a new plant or facility. After the facility is built, the reliability program is more of a backfit. However, that is probably when 98 percent of all reliability programs are developed.

The benefits of developing a reliability program as part of the design stage include the deliberation, to the extent possible, for allocating and devising ways to replace certain equipment without having to schedule an outage for accomplishing the maintenance work. Other benefits include the design of equipment so that it can be maintained without having to first remove half of a building to gain access to it. Additionally, provisions can be made for utilizing predictive maintenance techniques to the maximum level for monitoring equipment, thereby minimizing the need for intrusive maintenance.

Industries where the facility is a cruise ship, an aircraft, or a new power plant—or any new facility, for that matter—should be especially attuned to this issue because they have the opportunity to be proactive about establishing a reliability program.

Aircraft manufacturers, for example, are quite familiar with this. Usually, the competition among two or more manufacturers for selling the next-generation aircraft adopts a marketing strategy that addresses the ease of maintainability, the simplicity of accessibility, and the capability to perform online predictive maintenance, all of which results in longer flying times and less downtime in the hangar.

One day this fundamental understanding of reliability will be so commonplace that reliability programs in general will become more a part of the design phase than a retrofit. It is preferable to design equipment and facilities from the start in such a way that equipment failures do not pose safety or operational concerns that lead to unwanted consequences.

6.7 Is a Design Change Recommended?

The last column of the PM Task Worksheet, column G, includes the recommendation for whether a design change is warranted. As we discussed in Chapter 3, a design change is the exception rather than the rule because it is not that often that a condition-directed or a time-directed task to prevent failure, or a failure-finding task to identify when a failure has occurred, cannot be specified. However, there are those occasions when this is not possible and a design change is mandatory or else you accept the risk of failure.

Most tasks are relatively simple and inexpensive. They do not automatically migrate toward a complete overhaul. As we have learned, intrusive overhauls and replacements are the last choice in preventive maintenance. In Chapter 3, we analyzed the fuel oil pumps for the emergency diesel generator. We found that those pumps were potentially critical because of their hidden failure consequences. The optimum task in that instance was a failure-finding task to ensure that *both* pumps were operating via an inspection. The inspection was included as part of the procedure governing the diesel each time it was started, which was once every 30 days. A design change was not necessary.

6.8 Completing a Typical PM Task Worksheet

To see how the PM Task Worksheet is completed, let's look at a typical example. The PM Task Worksheet is shown in Figure 4.3. We will use the two-way valve GHI that we analyzed in Figures 5.4*a* through 5.4*g* as our example. On the PM Task Worksheet, we enter the component I.D. and description, and in column B, we enter the critical consequences of failure for valve GHI.

Component I.D. and description.
  Valve I.D.: GHI
  Two-Way Isolation Valve

What were the consequences of failure?
1.a. A power reduction.
1.c. A plant shutdown.

In column C of the PM Task Worksheet, we describe the dominant component failure modes associated with the two consequences of failure. *Each dominant failure mode that resulted in an unwanted consequence of failure must be analyzed.* This is entered in column C of the PM Worksheet.

Describe each dominant component failure mode.
1.a. Valve fails closed.
1.c. Valve fails to transfer to the emergency position.

In column D of the worksheet, we describe all of the possible credible failure causes for each dominant component failure mode. Remember, this is just a representative example for a typical valve type and depends on the exact type of valve installed in your facility; the causes of failure may be very different from the ones in this example. However, I have also intentionally used this example to show that the causes for each different failure mode could be similar. In this instance, the causes for the failure mode of "valve fails closed" and the failure mode of "valve fails to transfer to the emergency position" are similar. This is not unusual, and it is acceptable to specify the credible failure causes for 1.c. as "the same as for 1.a."

Describe the credible failure cause for each dominant failure mode.
1.a. (1) Valve binds
   (2) Bearing failure
   (3) Motor failure
   (4) Torque switch fails
   (5) Limit switch fails
   (6) "Other," depending on specific equipment type
1.c. (1) Valve binds
   (2) Bearing failure
   (3) Motor failure
   (4) Torque switch fails
   (5) Limit switch fails
   (6) "Other," depending on specific equipment type

In columns E and F of the worksheet, the applicable and effective PM tasks and the frequency and interval of the tasks are defined for each failure cause. There is such great diversity in this decision and selection process that I will not attempt to include any typical examples. The tasks and their periodicities are ascertained from the PM Task Logic Tree and Figure 6.5. As I mentioned earlier in this chapter, the tasks and their periodicities are determined by prudent technical decisions made by knowledgeable individuals. Finally, column E is where you can define whether a design change is recommended.

6.9 Institute Technical Restraints

You are probably thinking, "What exactly does this mean?" The message here is to avoid what repeatedly seems to happen: once the RCM program is established, new tasks get added with very little oversight or scrutiny. Seldom do they undergo the same rigors of analysis that the original program was based on. Before you know it, your PM program has grown by 20 percent to 30 percent. One way to avoid the buildup of *non*applicable and effective PM tasks is to implement a process that ensures that any new proposed additions to the program go through the same COFA logic analysis for their justification. If you do not institute some type of requirement for a technical justification before any additional task can be added to the program, you will find that your program will grow and slowly accumulate unnecessary work.

As we discuss in Chapter 8, new tasks will always be required as new issues develop, and you do not want to stifle that part of the process. However, you should not allow the process to get out of control, either. This is not always easy to do; probably the best way to manage this restraint would be to implement an approval process for new PMs and make it the administrative responsibility of the department in your organization that is accountable for preventive maintenance.

6.10 A Sampling Strategy

Now that we have learned how to recognize critical and potentially critical components and understand a little more about the PM tasks and the periodicities of those tasks, let's explore another area that is very much a part of real life in the world of maintenance and reliability but is seldom discussed.

What if you have just completed your RCM analysis and found that one of the most critical components you have is a major pump, for example. The pump could be quite large and very complex. There might be a half dozen similar pumps in your plant, all of which are critical and have been in service for 15 or more years and have never been overhauled before. Now, assume that it will cost several thousand dollars (maybe even several hundred thousand dollars) to accomplish the overhaul. Will you go to senior management and tell them that you just completed this RCM program and the results have determined that all these pumps should be overhauled? Suppose you have several dozen large motors in your plant that likewise are all critical, have been in operation for over 15 years, and have never had the stators rewound or the bearings replaced. Would you schedule all of them for overhaul at once? Probably not.

Even though you have strong suspicions that those components definitely need attention because pump seals, impellers, stator windings, and bearings don't last forever, even if you have been religiously performing the routine predictive tasks such as monitoring for excessive vibration, sampling and changing the oil, and so on, it would still not be an easy feat to convince senior management to buy into this grandiose overhaul plan—that is, unless your senior management is visionary, engineering-oriented, reliability-savvy, and willing to make bold decisions. Sadly, though, most senior management does not fit that mold. They would be especially hard to convince if none of these critical components has failed yet. Remember reading in Chapters 1 and 3 about running on luck? So what do you do?

The most prudent approach to this administrative problem is to convince senior management that you need to perform a "sampling" overhaul inspection of at least one of each type of equipment. Select the one in the most hostile environment and the one that has been in service the longest. The aim is to inspect the one that was operating under the most severe conditions. Once the sample component is removed, I recommend a very thorough inspection of all aspects including clearances, the condition of the stator windings, the condition of the bearings, the impeller wear, and any other precise measurements that will indicate its condition or degradation.

Depending on what you find, you should have a more technically based initiative to deal with the rest of that specific population. That way you will at least know if the equipment is in good enough condition to continue operating for many more years without concern. Or you may find that the entire component population is in imminent danger of failing in a short time. This is a real-life problem that has occurred in nuclear and other industries as well. Dealing with the remaining population of equipment then requires crisis management. Many facilities find themselves in a worst-case situation where a host of imminent failures are on the horizon. That is the price paid for the neglect of a premier preventive maintenance program.

Another reason for performing a sampling inspection is to confirm the validity of your predictive maintenance technologies. It is not uncommon for major failure mechanisms to go undetected on some equipment, even with a swarm of PdM tasks that are regularly accomplished. This, too, is a real-life problem that occurs all too often.

6.11 Common Mode Failures

Common mode failures are failures of a population of equipment all of which is subject to the same failure mode. As with the problem of an entire population of similar equipment failing within a similar time frame due to neglect of a preventive maintenance program, you must also be vigilant about the possibility of common mode failures occurring to a population of similar equipment. This is a situation to avoid.

Common mode failures have the potential to occur anytime you have a significant population of similar equipment. Examples include a facility with a large population of motor-operated valves or air-operated valves, all of which were manufactured by the same vendor, or only a few of each type of valve in dozens of your plants around the world. A typical commercial aircraft may have only one or two similar valves per plane, but there may be hundreds of that type of aircraft in the fleet.

Imagine what would happen if you discovered that a flaw existed during the assembly of a component, or an incorrect part was installed at the common shop that performed all of the overhauls, or a failure mechanism just manifested itself in a population of several dozen or even several hundred like components. You would definitely not be in a comfort zone, but it is not an uncommon situation in which to find yourself. When such situations occur, a mandatory bulletin or directive is usually issued by the FDA, NASA, the NRC, or the FAA, if you come under regulatory scrutiny, to inspect all suspect components within a very short time frame. Even if you aren't subject to regulatory scrutiny, you will come under economic scrutiny by senior management when they learn that several of the corporation's facilities are in jeopardy of an unscheduled and premature shutdown, perhaps for reasons that you cannot justify.

There is no absolute guarantee that this situation will never happen to you. However, there are actions that can be taken to minimize your exposure to such occurrences. It would be prudent to institute a sampling program that periodically inspects one or two of the components of the population to ensure that all is well *before* any problems rear their ugly heads and allow you very little time to react appropriately and efficiently with a planned course of action. A sampling program that performs an in-depth internal inspection of the sampled components can yield a very accurate prognosis for wear patterns, internal flaws, and other incipient failure mechanisms that may not be observable using PdM techniques.

6.12 Different Predictive Maintenance (PdM) Techniques

We have learned about condition-directed maintenance and how PdM, which is a subset of condition-directed maintenance, is the preferred PM task because it is mostly nonintrusive. As I mentioned in Chapter 3, PM tasks can be grouped together, and it is more efficient to handle them that way. Some of the newer methods employ templates that usually include an array of predictive techniques that are applied to a family of like components. The periodicity of the PdM tasks is different for each one depending on the environment, criticality classification, and operating conditions such as whether the component operates in continuous duty mode or cyclic duty. For example, a common template could be applied to all electric motors with a certain horsepower rating. This template would consist of several different PdM tasks including vibration analysis, oil sampling, MCSA, thermography, inspection, and so on. The time intervals for the PdM tasks, however, would be customized for each motor. Let's look at a few of the different types of PdM techniques.

6.12.1 Vibration monitoring and analysis

This application is used to detect bearing wear, an unbalanced condition, or other alignment problems mostly in rotating machinery. Vibration sensors can detect loose or cracked support mounts or support pads, bent or cracked shafts, and coupling problems. The software for vibration analysis enables trending of the vibration levels so that incipient failures can be monitored as they progressively get worse.

6.12.2 Acoustic monitoring

This application is used to detect internal and external leaks in all types of valves such as motor-operated valves, air-operated valves, manual valves, and check valves. In these cases, usually air or water leaks by or through a butterfly, diaphragm, or other internal part of a valve. This type of monitoring also detects leaks through heat exchanger tubes that are cracked.

6.12.3 Thermography or infrared monitoring

This is a very commonly used technique for finding "hot spots" using an infrared camera. Electrical connections that have become loose are easily detected by this technique. This application is also used to detect any parameter that may be running hotter than normally expected, such as relay coils and motor windings.

6.12.4 Oil sampling and analysis

This is a very common technique to detect incipient bearing or internal gear failures. A small amount of oil is drained from the unit and analyzed and inspected for wear particles and other contaminants such as water intrusion into the oil reservoir. Some of the latest advances in oil sampling include simple-to-use on-site equipment. Previously, oil samples were usually sent to an off-site laboratory for analysis.

6.12.5 X-ray or radiography inspection

This application uses X-rays to detect subsurface flaws in welds or other metallic parts such as casings or valve bodies. Extreme care must be taken when employing this technology to avoid harm to personnel from the X-rays.

6.12.6 Magnetic particle inspection

This application detects surface cracks in metallic parts by setting up a magnetic field around the part to be inspected.

6.12.7 Eddy current testing

Eddy current testing is very similar to magnetic particle inspection. It produces a magnetic field and an eddy current flow to detect surface flaws.

6.12.8 Ultrasonic testing

This method also detects flaws in metallic parts. It is similar to radiography, except it uses sound waves to detect the flaws.

6.12.9 Liquid penetrant

This technique uses a dye to detect surface cracks in pipes, welds, and other metallic parts. The dye is viewed under an ultraviolet light, and the crack becomes visible where the dye was absorbed into it.

6.12.10 Motor current signature analysis (MCSA)

This technique detects problems with motors, such as cracked rotor bars and some motor winding problems. A clamp-on probe analyzes motor current traces. Experience with MCSA has been somewhat mixed relative to the accuracy of its findings.

6.12.11 Boroscope inspections

This technique uses a boroscope to visually inspect internal parts of equipment that cannot be inspected externally. Of course, there must be points of entry to accept the boroscope for the inspection. This is commonly used for turbine inspections and internal inspections of large valves.

6.12.12 Diagnostics for motor-operated valves

The commonly used diagnostic technique for motor-operated valves is called MOVATS. This consists of a set of equipment that recognizes a pattern signature of the valve to measure a host of readings. These include motor current, torque switch settings, stem thrust, switch actuation, and the electrical condition of the motor.

6.12.13 Diagnostics for air-operated valves

This is a diagnostic test box used to detect diaphragm leakage, air-operated solenoid problems, and other internal problems that arise with air-operated valves.

Many of the PdM techniques described here are used in conjunction with other PdM tasks. For example, oil sampling and analysis, vibration monitoring, and thermography may all be employed on the same component. Using several different PdM techniques will usually detect an incipient failure, where using only one PdM task alone may not. New PdM technologies are becoming cutting-edge, and newer techniques are being developed every day. Continuous monitoring with some of these PdM applications, rather than the periodic use of these techniques, offers even greater potential for recognition of incipient problems.

6.13 Chapter Summary

This chapter has shown how a very important element of the RCM effort—selection of PM tasks—can be made straightforward and uncomplicated. Refer to Figures 6.6a and 6.6b, which reinforce the idea that classical RCM can indeed be made simple.

Let's summarize what we have covered in this chapter.

- There are three general categories of preventive maintenance tasks: condition-directed, time-directed, and failure-finding.
- There are a host of different types of tasks that fall into one of the three categories. They include overhauls, inspections, per-

formance tests, bench tests, replacements, disassemblies, cleaning, and a whole host of PdM tasks.

- Condition-directed maintenance is *not* run-to-failure.
- Condition-directed PMs are preferred over time-directed PMs because, as a rule, they are nonintrusive. Furthermore, the studies of Nowlan and Heap show that approximately 89 percent of all components fail randomly, so condition-directed PMs are more efficient because they allow the equipment to operate until the imminent end of their life.

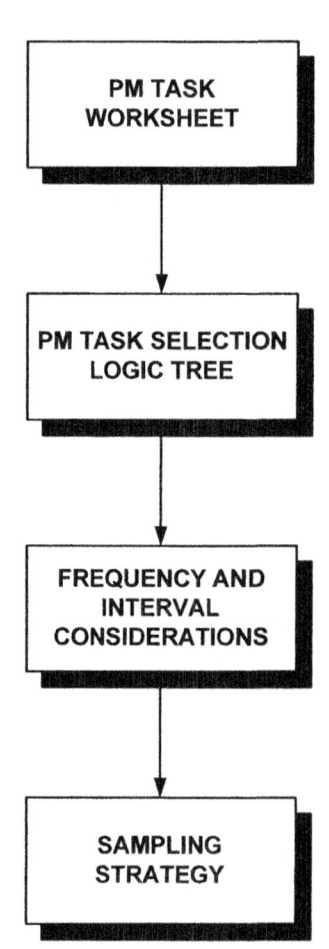

Figure 6.6(a) PM task selection logic made simple.

Describe the component failure mode

Describe the failure cause for each failure mode

Describe the PM task for each failure cause

Define the frequency and interval for each PM task

Describe if a design change is recommended

Figure 6.6(*b*) RCM logic sequence for specifying PM tasks.

- Predictive maintenance techniques are used to determine the condition of the equipment so that required overhaul or replacement can be planned and scheduled to preclude the occurrence of a functional failure.
- Time-directed PMs such as overhauls and replacements should be specified when an age-to-failure relationship is known. The studies by Nowlan and Heap show that only approximately 11 percent of all components exhibit an age relationship to failure.
- Failure-finding tasks are applicable to potentially critical components because potentially critical components are the result of hidden failures. Failure-finding tasks are not applicable to critical, commitment, or economic components because their failure consequences are immediate. Failure-finding preventive maintenance is also performed on safety systems and components that are not normally operating (they operate on demand).
- Operations, engineering, and other departments also perform PM tasks in addition to the maintenance department.
- There are usually several failure causes responsible for each failure mode.
- There are usually several PM tasks to prevent each failure cause. Selecting PM tasks requires prudent engineering judgment on the part of individuals who are knowledgeable about the equipment.
- The failure cause must be credible.
- The PM task must be applicable and effective.
- In the absence of an applicable and effective task, a *design change* may be required.
- Determining the optimum frequency and interval is based on many different considerations and requires prudent engineering judgment on the part of individuals who are knowledgeable about the equipment.
- The optimum time to establish a reliability program is during the design stage of the facility.

- A sampling strategy is always prudent and sometimes necessary.
- PdM tasks include vibration monitoring, acoustic monitoring, thermography, oil sampling, X-ray inspection, magnetic particle inspection, eddy current testing, ultrasonic testing, liquid penetrant, MCSA, boroscope inspections, motor-operated valve diagnostics, and air-operated valve diagnostics.

There is one category of equipment we have not discussed to this point. How do you handle *instruments?* Chapter 7 addresses this topic.

Chapter 7

RCM for Instruments

Instruments are an important part of your maintenance strategy, and they constitute an area that needs to be addressed in order to ensure their reliability. Instruments are normally segregated from other types of components. Although they do have a distinct equipment I.D., they are usually in a class type of their own. In most industries there are special technicians who provide for the calibration, maintenance, replacement, and repair of instruments. Different industries use different conventions for terminology applied to instruments. For example, instruments can function to control another component, switch something on or off, or provide an alarm; they can provide a reading of a certain parameter such as temperature or pressure; or they can monitor an electrical current or voltage. In many industries, black boxes are considered instruments. In some industries, instruments are also identified as instrument "loops." An instrument loop might consist of a transmitter, a sensor, an indicator, and a signal conditioner.

Instruments generally receive very little attention in terms of justifying when a calibration interval can be extended or when it should be reduced. Likewise, very little attention is given to the tolerance criterion that constitutes a calibration failure.

This chapter offers some guidance in regard to the preventive maintenance of instruments that provide only an indicator reading. Functional instrument preventive maintenance is governed by the RCM process. Nonfunctional instruments do not employ a specific process, but there are some logical decisions that can be made about their preventive maintenance activities.

7.1 Instrument Categories

I classify instruments into two categories: those instruments that provide a *function* and those instruments that provide only an *indication,* usually from a gage or a monitor. This second category includes all instruments that do *not* provide a specific function. Many times a reading on a gage causes an action to be taken that requires the initiation of a function, but that action is taken by a person, not the instrument.

If an instrument provides a function, it is analyzed in the COFA Worksheet and the PM Task Worksheet along with all other components. If it does not provide a function, it is analyzed in accordance with the Instrument Logic Tree, explained later in this chapter.

An instrument code of FI, designating a *functional instrument,* is assigned to those instruments that provide a function. If the instrument provides a function, it is no different from any other functional component—it encompasses a function, a functional failure, a failure mode, and a consequence of failure. It will also have a cause of failure and an applicable and effective PM to address each cause of failure.

An instrument code of II, designating an *indication instrument,* is assigned to those instruments that provide only an indication readout, typically via a gage or a monitor. The majority of instruments are in the II group.

The component classification of FIs is governed by the COFA, the Potentially Critical Guideline, and the Economically Significant Guideline. Instruments designated as II are analyzed in the Instrument Logic Tree, as shown in Figure 7.1. If an indication instrument is sufficiently important that a certain indication readout requires that personnel actions be taken, the instrument cannot be considered for a run-to-failure classification. However, if its indication readout does not require any immediate intervention and it is found to have redundancy, then it may be considered for run-to-failure status. The typical preventive maintenance activity for IIs is a calibration at a given periodicity.

RCM for Instruments 183

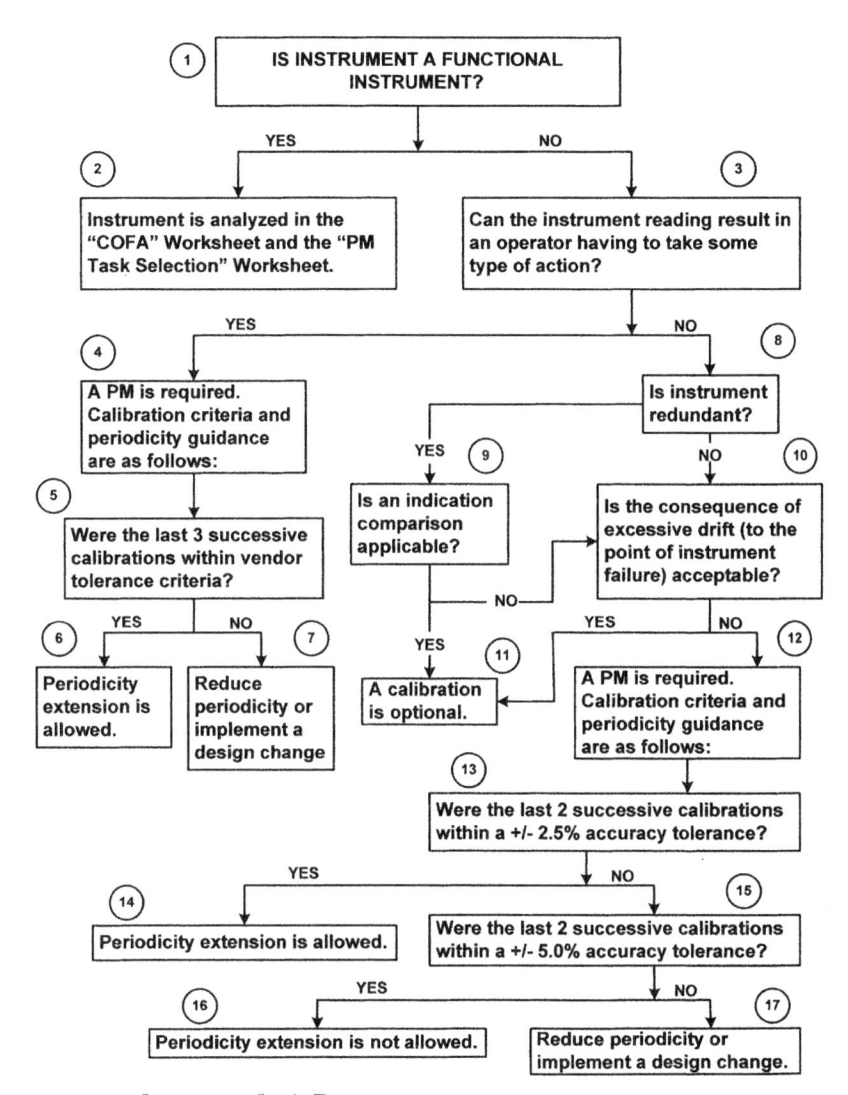

Figure 7.1 Instrument Logic Tree.

7.2 Instrument Design Tolerance Criteria

Most maintenance calibration programs for indication instruments revert to their accuracy design tolerances of a nominal +/− 0.25 percent or something similar, which is the criterion established for brand-new instruments leaving the vendor. Instruments are subject to drift. Maintenance calibration history data is usually the sole data source used as justification for changing calibration periodicities. Using the +/–0.25 percent tolerance levels almost ensures that most routine gages will not meet this criterion when they are normally calibrated, and the calibration test will thereby be declared "failed" when the instrument is calibrated.

In many instances this is too restrictive to allow sufficient flexibility in adjusting calibration periodicities, so the Instrument Logic Tree allows for more prudent accuracy tolerances to be used when applicable. A razor blade, for example, has a factory-specified tolerance. After you use that razor blade only one time, it no longer meets the restrictive factory design tolerance, but as we know, the razor blade is still usable until it is determined to be out of tolerance for normal use.

I have been very deliberate and specific about when these more realistic accuracy tolerances can be used. If you should elect to use these relaxed tolerance criteria, I suggest that you obtain approval from your management to do so. I have included them only as anecdotal information based on experience. However, it should be understood that whenever any instrument is recalibrated for any reason, it is always recalibrated using the original design accuracy tolerances.

The as-found calibration criteria used to determine a calibration failure for functional instruments (FI) are always in accordance with the vendor design tolerance standard criteria. Relaxed tolerance criteria for FIs are not permitted.

Figure 7.1 very deliberately specifies the number of previous successive calibration readings that are required to have been found "acceptable" in order to justify making any changes to the calibration periodicity. For nonfunctional instruments whose readings could result in operator actions being taken, three previous successive calibrations must have been in accordance with vendor tolerance criteria in order to justify a periodicity extension. If the last three consecutive readings did not meet vendor criteria, a *reduction* of the periodicity or a design change should be explored.

For nonfunctional instruments whose readings would not result in an operator having to take some type of action, the tolerance criteria are relaxed, and there is the possibility that a calibration may not be required.

7.3 The Instrument Logic Tree

Each block of the Instrument Logic Tree is explained in sequence.

7.3.1 Block 1: Is the instrument a functional instrument?

Starting with block 1, first determine whether the instrument is a functional instrument (FI) or a nonfunctional, indication-only instrument (II). To be a functional instrument, it must provide a function; for example, it provides an automatic trip function, a control function, an alarm function, or an interlock function. If it is a functional instrument, the logic proceeds to block 2.

7.3.2 Block 2: Instrument is analyzed in the COFA worksheet and the PM task selection worksheet.

If the instrument is a functional instrument, it is analyzed in the COFA Worksheet and the PM Task Selection Worksheet like all other components. If it is not a functional instrument, it proceeds along a logic path beginning with block 3.

7.3.3 Block 3: Can the instrument reading result in an operator having to initiate some kind of action?

Block 3 asks the question "Can the instrument reading result in the operator having to take some type of action?" This means that even though it is a nonfunctional instrument, the information readout from that instrument may result in operations personnel having to initiate some type of action or manually perform a certain function. This function might be a requirement to shut off a pump if a pressure reading gets too high or write a corrective maintenance order after observing an instrument reading indicating that a component requires corrective maintenance. This includes instruments used during operator rounds to verify the functional operability of components whose failure can result in an unwanted consequence.

It is not uncommon that some personnel action is required when an instrument reading indicates a certain parameter limit. Sometimes a regulatory requirement or some other required commitment depends on observing an instrument reading to determine whether a specific parameter limit has been reached. This includes virtually all control room instruments.

The purpose of asking this question is so that a prudent maintenance strategy is applied to a nonfunctional but important instrument, preventing it from being a candidate for run-to-failure. A *yes* answer to block 3 takes us to block 4. A *no* answer takes us to block 8.

7.3.4 Block 4: A PM is required. Calibration criteria and periodicity guidance are as follows.

This identifies that a PM is required. If there was no previously existing PM, it should be added. Usually, this entails instituting a PM to calibrate the instrument. Block 4 also includes specific calibration criteria and periodicity guidance. Note that this guidance is more stringent than the logic that would follow if the answer to block 3 was *no*. This takes us to block 5.

7.3.5 Block 5: Were the last three successive calibrations within vendor tolerance criteria?

Block 5 asks whether the last three successive calibrations were within the vendor accuracy tolerance. This is the nominal +/−0.25 percent tolerance. Since this block includes relatively important nonfunctional instruments, a history of the last three successive readings must be shown to have been within vendor tolerance criteria. If the answer is *yes,* proceed to block 6. If the answer is *no,* proceed to block 7.

7.3.6 Block 6: Periodicity extension is allowed.

If the last three successive calibrations were within the vendor tolerance limits, an extension of the calibration periodicity is allowed.

7.3.7 Block 7: Reduce periodicity or implement a design change.

If the last three successive calibrations were not within vendor tolerance limits, an extension of the calibration periodicity is not allowed. A review of calibration records should be made to determine whether a reduction of the periodicity or a design change should be explored.

7.3.8 Block 8: Is the instrument redundant?

This question is to determine whether the instrument is redundant. Is there another independent instrument not prone to common mode failure associated with this instrument that measures the same parameter or a comparable parameter? Oftentimes there are several instruments measuring the same parameter.

7.3.9 Block 9: Is an indication comparison applicable?

This question asks whether or not one may reasonably expect comparisons to be periodically made such that excessive drift of one instrument would be noticed and corrective action could be taken. To be considered for a comparison, the comparison readings must be part of a documented procedure.

Instances where this would be a reasonable expectation include the following:

- Both instruments are monitored and are near each other, where an operator would easily see both.
- Both instruments are recorded on a log taken or reviewed by the same individual.

- Both instruments are monitored by a computer programmed to recognize a disparity between readings.

Instances where this would not be considered a reasonable expectation include the following:

- The redundant instrument is not normally accessible, that is, it is located in an area that is not readily accessible.
- The redundant instrument is not comparable in terms of indication readout units.
- A computer monitors both points but does not compare them.

7.3.10 Block 10: Is the consequence of excessive drift (to the point of instrument failure) acceptable?

In general, consider the economic and operational consequences of instrument failure or drift. Evaluate the ability of the specified PM to prevent the failure or drift. Consider the cost or consequences of equipment failures that are monitored by the respective instrument.

If problems that would be indicated by this instrument would most likely show up on other instruments as well and if the cost of the failure is low and the cost of the PM is low, then the PM is optional.

On the other hand, if the instrument fails or drifts to the point of failure, resulting in a condition that could last for a long time without appropriate indication, excessive drift is *not* acceptable. If the cost of failure of the equipment being monitored is high, excessive drift would similarly *not* be acceptable.

This block considers instruments that are *not* significant to either plant safety, operability, or economics. As such, if the consequence of excessive drift is acceptable, but not to the point of instrument failure, then an allowable drift tolerance of up to +/−5 percent is considered acceptable. If drift to the point of instrument failure is acceptable and no operator evolutions are based on the instrument readings, then a calibration PM is optional.

7.3.11 Block 11: A calibration PM is optional.

This block is self-explanatory.

7.3.12 Block 12: A PM is required. Calibration criteria and periodicity guidance are as follows.

This identifies that a PM is required. If there was no previously existing PM, it should be added. Usually, this entails instituting a PM to calibrate the instrument. Block 12 also includes specific calibration criteria and periodicity guidance. Note that this guidance is less stringent than the logic that would follow if the answer to block 3 was *yes*.

7.3.13 Block 13: Were the last two successive calibrations within a +/−2.5 percent accuracy tolerance?

This block is applicable only to instruments *not* considered significant to plant safety, operability, or economics and when excessive drift up to 2.5 percent is acceptable. If the last two successive calibration readings were within +/−2.5 percent, then the calibration is considered to be satisfactory and a periodicity extension is allowed. If the drift was not within +/−2.5 percent, block 15 asks whether the drift was within a +/−5 percent accuracy tolerance.

7.3.14 Block 14: Periodicity extension is allowed.

An increase in the PM periodicity is allowed.

7.3.15 Block 15: Were the last two successive calibrations within a +/−5.0 percent accuracy tolerance?

This block is also applicable only to instruments *not* considered significant to plant safety, operability, or economics and whose calibration drift history is greater than 2.5 percent but less than or equal to 5 percent. If the last two successive calibration readings were between 2.5 percent and 5 percent, then the existing periodicity is acceptable and should not be changed. If the last two successive calibration readings were greater than 5 percent, a decrease in the PM periodicity or a design change should be considered.

7.3.16 Block 16: Periodicity extension is not allowed.

The existing PM periodicity is acceptable and an increase in the periodicity is not allowed.

7.3.17 Block 17: Reduce periodicity or implement a design change.

A reduction of the PM periodicity or a design change should be explored.

7.4 Chapter Summary

A review of this chapter includes the following key points:

- Instruments are usually in a class type of their own.
- There are two fundamental categories of instruments: *functional instruments* (FI) and *indication-only instruments* (II). Instruments may also be grouped together to form a loop whereby the entire loop is considered to be the instrument.
- Functional instruments provide a function and are no different from any other functional component. They have a function, a functional failure, a failure mode, and a consequence of failure associated with them. They also have a cause of failure and an applicable and effective PM addressing each cause of failure.
- Functional instruments are analyzed in the COFA, the Potentially Critical Guideline, and the Economically Significant Guideline.

- Indication instruments provide only an indication readout, typically via a gage or a monitor. The majority of instruments are in this group.
- Indication instruments are analyzed in the Instrument Logic Tree.
- Instruments are subject to drift.
- Calibration criteria for indication instruments usually revert to the design tolerance of +/–0.25 percent. For many indication instruments, this is too restrictive.
- The Instrument Logic Tree provides for relaxed tolerances in a prudent manner depending on whether the instrument reading can result in an operator having to take some kind of action.
- The Instrument Logic Tree provides guidance for either extending a calibration PM periodicity, reducing a periodicity, or implementing a design change.
- Previous calibration histories are required to justify periodicity changes.
- If indication instruments are redundant and comparison readings can be obtained, a calibration PM is optional.
- When any instrument, FI or II, is calibrated during its PM activity, it is always calibrated to its design tolerance.

Chapter 8

The RCM Living Program

What is a *living program?* Like life itself, if a program is not living and evolving, it is either dormant or deceased. A living program is a program that continues to grow, evolve, change, and adjust. It continues to breathe and adapt to alterations to its original makeup. Not unlike the fields of medicine and science that are constantly evolving, a maintenance program also must evolve. If a maintenance program remains stagnant, it no longer offers the benefits of its intended objective. Many people believe that once they implement their RCM program, they are finished forever. They allow the program to go into a dormant mode. What if medical science remained dormant? We would not have the modern medical wonders of today, which include the entire field of genetics, being on the verge of a breakthrough on curing Alzheimer's disease, laser eye surgery, and new pharmaceuticals that are being developed almost every day.

In Chapter 5, when we defined the asset reliability strategy, I made the statement that RCM is the *driver* for a corporate reliability program. This becomes more evident as industry faces greater worldwide competition. This, in turn, spurs plants and factories to be more diligent about finding more effective and efficient ways of doing business and maintaining profitability. As we also know, the maintenance budget for any plant or facility constitutes a significantly large portion of the total outlay. Thus, maintenance has a significant role in the bottom line of the corporation. As worldwide competition continues to grow, the corporation must change and adapt to that competition, and the maintenance program of the corporation must also evolve and stay on the cutting edge of new maintenance technologies, continuously adjusting its PM program to be as efficient as possible.

You have learned how to develop a premier RCM program based on the data that you had at the time of implementation. This is especially true for the periodicities you prescribed for the PM tasks. However, everything is subject to change. These changes can come about as a result of new failure modes that were nonexistent at implementation but have become apparent since that time. New changes in maintenance methods may have been sufficiently beneficial to allow periodicities to be extended. On the other hand, some changes in equipment parameters may result in the need to reduce periodicities to maintain reliability. Modifications to the plant, modifications to equipment, changes in operating characteristics, new PdM technologies, which are appearing everyday—all these factors contribute to the need to periodically review and update your maintenance program.

The *results* of the RCM maintenance program are not set in cement; rather, they are set in malleable clay. What is set in concrete is the *RCM decision logic* that determines which equipment is required to have a preventive maintenance strategy and the *PM Task Selection Logic* that determines the different categories of preventive maintenance. The RCM methodology does not change, but the results of the analysis certainly will.

8.1 A Model for an RCM Living Program

This chapter explains some of the methods to help you keep your RCM program current and accurate. Like a car that periodically requires a tune-up to keep it running efficiently and reliably, a maintenance program also requires a periodic tune-up. The difference, however, is that the *periodicity* for the maintenance program tune-up presents itself on an almost daily basis. It becomes an automatic tune-up—self-adjusting based on the evaluation of completed PM tasks; or with new input from the various vendors, input from the industry at large, new requirements from the applicable regulators; or by the acquisition of newly discovered operating and design characteristics of the equipment itself.

For example, I will explain methods for making changes to PM periodicities based on the feedback of the craft personnel performing the tasks. I have also included corrective maintenance (CM) as an element of the process. Additionally, results and feedback from a monitoring and trending program have been included as an element of the living program. The monitoring and trending program for plant performance is discussed in detail in Chapter 9.

It is essential for a living program to have some logic embedded in the process; otherwise, you will oscillate the PM program

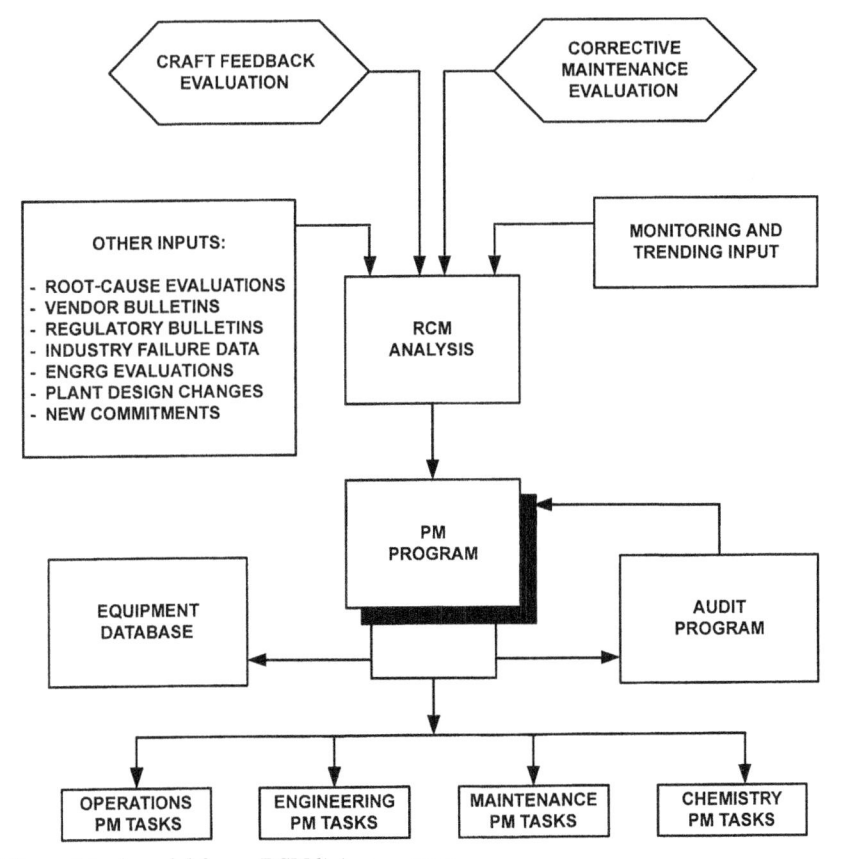

Figure 8.1 A model for an RCM living program.

in a back-and-forth manner by extending periodicities and then having to reduce them because you extended them too far. This wastes time and only results in confusion for everyone involved. Therefore, I have inserted some "intelligence" into the decision processes to avert the possibility of overshooting and thus having to move a periodicity back and forth. This is all part of the craft feedback element of the living program.

Figure 8.1 shows a model process for an RCM living program. Notice how the logic flows. All inputs feed directly into the RCM analysis. The craft feedback element and the corrective maintenance evaluation element are the main inputs. The other inputs and the monitoring and trending input also feed directly into the RCM engine. It is the RCM engine that drives the PM program. Extending from the PM program is the equipment database and an internal audit program. The PM program then becomes the driver for the different organizations that perform PM tasks. As noted in Chapter 6, maintenance is not the only organization performing PM activities.

Let's examine each element in detail.

8.1.1 The craft feedback evaluation element

This is perhaps one of the most significant elements of the living program. There are many reasons why this element is important. A primary reason is that it provides direct feedback from the craft personnel performing the PM tasks that you established. This element verifies that the PM tasks are indeed accurate and correct, that they are scheduled at the right periodicity, and that they reflect the scope of work that is needed. When your PM frequency and intervals were first established, they were based on many factors. The feedback from the craft either validates that those factors were accurate or presents a set of results that provide a basis for changing the originally selected periodicity or scope of work.

The second reason for the importance of this element is that the craft personnel themselves are involved in the process. Chapter 2 mentioned the importance of having the craft personnel involved in the RCM effort since they are the ultimate emissaries of the program. There probably is no quicker way to throw a wet blanket over all of your efforts than to specify time-directed PM tasks to be performed by craft personnel who can see that the equipment does not yet require the overhaul or replacement. This can make them think that their work efforts are being wasted—and they are correct to think so. If the craft personnel do not believe in the results of your efforts, it won't be long before senior management begins to develop the same suspicious inquisitiveness.

This is really a matter of the craft personnel inheriting pride of ownership in the program. Chapter 2 also stated that there is no monopoly on plant knowledge. Craft personnel are an extremely valuable source of information, and you will find them ready allies if you bring them into the process.

Experience has shown that the introduction of most new comprehensive programs that affect a host of different stakeholders achieve initial success. Nevertheless, if the program is force-fed to a major segment of the stakeholders, such as the working craft personnel, and they have no venue for making constructive suggestions, it won't be long before that success diminishes. Remember, the RCM effort may not reach 100 percent accuracy during the initial development process, but your program should strive for that 100 percent goal once it has been implemented, based on the feedback from the living program elements.

The craft feedback element can take on as much sophistication as you desire, or it can remain relatively simple and still be effective. It can be accomplished with simple forms, or it can be embedded in the software as part of your CMMS system, which was discussed in Chapter 4. For relatively large facilities with a sizable equipment population, a computerized software process is highly recommended. For relatively smaller facilities, a simple form may be all that is needed. Only you have the experience and the knowledge to make that decision. I will outline a Craft Feedback methodology and logic that can be implemented either through the use of simple forms or via a more comprehensive software program. You should see fit to modify, add to, or even delete any area in order to achieve a better outcome for your specific situation.

In the craft feedback element, each PM task allows for the expert opinion and best judgment of the craftsperson or technician performing the work to determine the appropriateness of the task itself. This includes the work scope of the task, the task periodicity, and whether there is a recommendation for some other modification or adjustment to the task. Craft personnel may also make comments regarding possible design changes to the equipment based on their experience. The scheduled PM should have a data field for the craftsperson or technician to enter a description of their judgment about the PM. Figure 8.2 offers a guide for training craft personnel on the relative condition of the PM when they perform a given task. It is strongly recommended that as part of this feedback element, you establish

---

OVERVIEW: The craft feedback grading categories are designed to utilize the professional opinion of the crafts to validate the existing PM for accuracy, thoroughness of the work scope, and appropriateness of the periodicity, or to justify other changes to the PM as applicable.

The craft feedback is not intended to describe the condition of the equipment, only the PMs associated with the equipment. Comments can be made, however, in regard to the equipment condition and noted on the feedback forms.

Category grade

5 *Good:* The PM condition is comparable to the same condition as if the PM was just worked. It is like new. Evaluate increasing the periodicity.

4 *Above average:* The PM condition is between good (5) and average (3). There is very minor degradation.

3 *Average:* The PM condition is adequate to allow the component to perform its function. Degradation is normal and is as expected. The PM is being performed at the correct periodicity.

2 *Below average:* The PM condition is between poor (1) and average (3). There is more degradation than expected.

1 *Poor:* The PM condition reveals that immediate attention is required. It is at a point where the function of the component has significantly deteriorated. Evaluate reducing the periodicity.

Figure 8.2 Craft feedback categories.

some type of communication, either directly or via e-mail or some other method of communication, to *provide feedback to the craftsperson or technician that their input has been received and duly reviewed.* Merely requesting one-way feedback without a return response detracts from the benefits you are trying to achieve. My own experience as well as industry experience affirm the importance of providing responses when feedback has been requested and received.

There are five craft feedback categories: good (5), above average (4), average (3), below average (2), and poor (1). *The feedback is based on the condition of the PM task and not on the overall condition of the equipment.* The overall condition of the equipment is important but is not part of the PM grading process. However, any pertinent craft comments about the equipment itself should be encouraged.

There should be adequate means for the craftsperson or technician to document their task category grade and any other comments, either on a hard-copy form or as data input to a software program. Keep in mind that the category grade feedback is for the PM task.

For example, if a PM requires the inspection and replacement of a filter and it was found that the filter was like new when the task was performed, a category grade of 5 would be chosen. Even if the component of which the filter was a part was in a failed state, the PM to inspect and change the filter would still be a 5. Conversely, if the filter was found to be completely clogged and unable to pass the required fluid flow, a category grade of 1 would be chosen.

The goal is to achieve a category grade of 3, which is average. You don't want the filter sparklingly clean, and you don't want it almost completely clogged. It should show some normal amount of degradation or debris buildup commensurate with your expectations, and it should enable the successful function of the respective component and its system, which indicates that it is being replaced at the appropriate periodicity. As for the failed component to which the filter was attached, a *comment* about that failed condition should be entered by the craftsperson.

The categories of below average and above average are included so that shades of degradation, or the absence of significant degradation, can be quantified as accurately as possible. If there were categories only for good, average, and poor, there would be insufficient opportunity for decision flexibility.

The feedback data only identifies the need to *commence* an evaluation. Each feedback data sheet should not in itself determine that there should be a change in a PM; each data input should be viewed as a flag for consideration by the evaluator of the data and should not automatically trigger a change. An evaluators checklist is shown in Figure 8.3*a* for category grades 1 and 2 and Figure 8.3*b* for category grades 4 and 5. I have added some intelligence to the decision process to account for other influences that can affect the category grade. For example, a grade of 4 or 5 may have been documented because corrective maintenance that included changing the filter was recently performed on the equipment. On the other hand, a grade of 1 or 2 may be documented because the PM was considerably overdue and it was not performed until long after its scheduled periodicity, which would obviously cause the filter to accumulate added debris.

Therefore, I recommend that at least two and possibly even three or four successive similar grades be documented before changes are made to the program. This way, an organized iterative process will be established to define the optimum PM frequency and interval based on the evaluators' judgment.

Also included in the evaluation checklist is a question to verify that the grade is relative to the PM task and not the overall condition of the equipment. Another question asks whether this was a random occurrence. For example, someone might accidentally have caused previous damage to the part by inadvertently stepping on it. Another question asks whether the grade relates to other, *identical* components. Credit for other, identical PMs can be used as successive PM grades for compiling the data accordingly to support a decision.

Here are some other factors for the evaluator to consider so that the program does not unnecessarily fluctuate:

- A grade of 3 (the average expected value) will not trigger an evaluation, since that task is being performed at the optimum periodicity.

```
      CRAFT FEEDBACK GRADE: _____

      EQUIPMENT I.D.: _____
```

PM NUMBER: _____

PM DESCRIPTION: _____

	YES	NO
1. DOES THE GRADE GIVEN RELATE TO THE PM TASK?	☐	☐
2. IS THE GRADE THE RESULT OF AN OVERDUE PM?	☐	☐
3. IS THERE A COMPONENT FAILURE HISTORY ASSOCIATED WITH THIS PM?	☐	☐
4. DOES THIS GRADE RELATE TO OTHER IDENTICAL COMPONENTS?	☐	☐
5. IS THIS A RANDOM OCCURRENCE?	☐	☐

RECOMMENDED ACTIONS:

	YES	NO
1. CHANGE PM WORK SCOPE?	☐	☐
2. CHANGE PM PERIODICITY?	☐	☐
3. IS A DESIGN CHANGE RECOMMENDED?	☐	☐

COMMENTS:_____

EVALUATOR: _____

DATE: _____

Figure 8.3(a) Evaluators checklist for craft feedback category grades (1) and (2).

- If you have more than one plant or more than one operating unit and they are virtually identical, the equipment feedback on one unit can be used to justify changes to the identical equipment in the other unit(s) but *only when it has been verified that the installation, environment, design, and operating conditions are identical.*

```
┌─────────────────────────────────────┐
│    CRAFT FEEDBACK GRADE: _____    │
└─────────────────────────────────────┘

┌─────────────────────────────────────┐
│    EQUIPMENT I.D.: _____      │
└─────────────────────────────────────┘
```

PM NUMBER: _____

PM DESCRIPTION: _____

	YES	NO
1. DOES THE GRADE GIVEN RELATE TO THE PM TASK?	☐	☐
2. IS THE GRADE THE RESULT OF PERFORMING THE PM EARLY?	☐	☐
3. WAS THERE A RECENT CORRECTIVE MAINTENANCE (CM) ACTIVITY SINCE THIS PM WAS LAST PERFORMED?	☐	☐
4. DOES THIS GRADE RELATE TO OTHER IDENTICAL COMPONENTS?	☐	☐
5. IS THIS A RANDOM OCCURRENCE?	☐	☐

RECOMMENDED ACTIONS:

	YES	NO
1. CHANGE PM WORK SCOPE?	☐	☐
2. CHANGE PM PERIODICITY?	☐	☐

COMMENTS: _____

EVALUATOR: _____

DATE: _____

Figure 8.3(*b*) Evaluators checklist for craft feedback category grades (4) and (5).

- If you have more than one identical component in the same system and it is the same make and model in the same environment (such as four identical circulating water pumps in the same system in the same environment), the data analyzed can be applied for all four identical pumps. *Note: Similar makes and models of equipment used in different systems in different parts of the plant and providing different functions are* not *con-*

sidered identical, and the feedback data from one cannot be applied to the other.

- Many facilities will use a worklist for their PMs so that a separate PM is not planned and scheduled four separate times, for example. In the case of the four circulating water pumps, the pumps are identical, they are in the same location, they function the same, and the filters on all four pumps are identical. Therefore, instead of issuing four separate PM tickets, only one PM is planned and it includes cleaning all four pump filters. Sometimes other work scope is scheduled on that same ticket—for example, changing the gearbox oil. If you should determine that the filter periodicity can be extended but the oil change cannot, then you have a choice: you can schedule the filters on a different ticket to capture the benefit of the extended periodicity, or it may be more appropriate to keep the same periodicity and perform the filter maintenance and the oil change together on the same ticket. The decision is yours.

- What about weekly, monthly, and quarterly PMs? These are performed so often that it should not take long to verify their optimum periodicity. Once that has been determined, you may wish to evaluate whether you want to continue to include the weekly, monthly, and quarterly PMs in your feedback program, if only to reduce the number of data forms you will be receiving. This will free up your time to concentrate on reviewing and analyzing the other data inputs for the rest of your components.

8.1.2 The corrective maintenance (CM) evaluation element

The preventive maintenance program is continually optimized by comparing prescribed preventive maintenance tasks against actual corrective maintenance (CM) experience in order to effect changes to the program as deemed appropriate.

This element of the living program evaluates corrective maintenance activities to review for the discovery of failure modes that may not have been identified in the original analysis or that were not apparent at the time of the original analysis. It also reviews CM histories to determine whether new PMs should be implemented or changes should be made to the scope of existing PMs. If a specific component has a high CM history that can be addressed by a PM, it might be that the component was overlooked as an economic component and should have been captured in the Economically Significant Guideline.

The CM evaluation could result in adding PMs, deleting PMs, increasing or decreasing the PM work scope, extending or reducing the PM periodicities, or determining that the plant can tolerate the corrective maintenance and no further action is required. The CM evaluation is a very important input to the living program to ensure that the preventive maintenance program is up to date and as efficient as possible.

If a critical, potentially critical, commitment, or economic component continues to have a history of pertinent corrective maintenance activity, it could be an indicator of any of the following conditions:

- The PMs for those components are deficient in their scope of work.
- The periodicities of the PMs for those components could be incorrect.
- The PMs are not being performed when scheduled.
- Deficiencies in the work control process exist relative to planning and scheduling the work.
- Parts availability is a problem.
- A design change may be necessary.

Of course, the PM tasks must have had an opportunity to be in place for at least one complete scheduled cycle. Corrective maintenance can occur if the PMs have not yet had time to be embedded in the program.

As I mentioned in Chapter 5, a CM count by itself is *not* an indicator of a deficient PM program. Some RCM books will automatically classify a component as critical if it has a large number of CMs. That is an incorrect indicator. There are a multitude of CMs generated to correct very minor deficiencies that do not threaten the operability or functionality of the component. Examples include minor nicks or scratches, minor packing adjustments, and missing I.D. tags. Counting these totally irrelevant CMs and using them to justify a preventive maintenance program decision is neither prudent nor cost effective.

The types of CMs that must be included in a preventive maintenance program decision process include those CMs that were generated due to a *failure or significant degradation* of a critical, potentially critical, commitment, or economic component.

The threshold for generating a CM varies depending on your industry and the type of plant or facility. Some industries do not allow any work to be performed that was not specifically included in the PM work scope. If the PM was to inspect a specific piece part and the part next to it was in need of repair, a new CM would be required for the piece part that was not included in the work scope. Some industries allow more flexibility about the additional work that can be performed based on the knowledge and experience of the craft personnel performing the PM. Understanding this supports the disavowal of another generally accepted fallacy: that a metric included in your PM program should be a PM/CM ratio. Any directive for a specific number of CMs that should be expected for each PM is questionable at best. Most reliability experts have abandoned this mistaken correlation long ago.

Another issue that is central to CMs is how they relate to predictive maintenance. As we learned, PdM is the task of choice for the PM program. But what happens when a PdM task finds a motor bearing vibrating excessively that requires immediate replacement? Is the motor replacement included as part of the PdM task (i.e., a condition-directed PM), or is a CM generated to replace the motor once the PdM task has discovered the excessive vibration? Chapter 10 delves deeper into this question.

8.1.3 The "other inputs" element

A living program is always affected by various inputs that result from a multitude of sources. Each one of these inputs can affect the PM program. Some of the more common ones are as follows.

8.1.3.1 Root-cause evaluations Depending on the industry, you may or may not be familiar with root-cause evaluations. In most regulated industries such as airlines and nuclear power, a component failure that resulted in a significant failure consequence entails a formal root-cause evaluation to not only determine the cause of the failure but also ascertain which barriers were insufficient to have allowed the failure to occur. Some of the factors that are considered include: What human-factor deficiencies existed? What procedural inadequacies existed? Was there a management culture issue that led to the failure? Has this type of failure occurred before? Has a similar failure occurred within the industry? The main objective of a root-cause evaluation is to prevent that occurrence from ever happening again. Several recommendations usually result from a root-cause evaluation. Invariably one of the recommendations includes some type of corrective action relative to the preventive maintenance program.

Some industries that are not regulated may nevertheless be familiar with this rigorous type of evaluation and are aware of the corrective actions that result. Industries not familiar with root-cause evaluations may wish to learn more about the value of implementing this strategy.

8.1.3.2 Vendor bulletins Vendor bulletins are another source of information that oftentimes result in a change to the preventive maintenance program. Vendor manual recommendations for preventive maintenance tasks to be performed on their equipment are obviously going to be ultraconservative. The vendors want to minimize their warranty obligations, and the more spare parts you order, certainly the happier their financial people will be. As noted in Chapter 6, vendor recommendations should be only one of many considerations for establishing your PM program. Routine vendor-recommended tasks should always be tempered by your own actual in-house experience.

Vendor bulletins are another issue. Quite often, a design deficiency or some other type of manufacturing flaw is reported to a vendor, and they are required to take action accordingly. A vendor bulletin should be taken more seriously than the routine vendor manual PM recommendations. You may need to take rather quick corrective action on equipment supplied by a vendor who has issued a bulletin delineating a design deficiency.

8.1.3.3 Regulatory bulletins Industries governed by regulatory entities often receive regulatory bulletins. These bulletins have a very high priority of importance. A regulatory agency frequently issues a bulletin—sometimes in the form of an information notice—when something needs immediate attention. This might be a new requirement to perform a specific inspection, or a certain part replacement might be required. This is quite common in commercial aviation when a design problem has been discovered by the airframe manufacturer or the FAA. In most instances when a regulatory bulletin or information notice is published, it affects some aspect of the maintenance program.

Bulletins and information notices can be issued by OSHA, EPA, NASA, NRC, FDA, or any regulatory agency. It is not uncommon that a totally new preventive maintenance program addition becomes a permanent requirement as the result of a bulletin or information notice. If the component under scrutiny has not already been identified as critical or potentially critical, it will become a commitment component.

8.1.3.4 Industry failure data This is another source of input to the living program. As previously mentioned, many industries share their failure data with each other. Even in a competitive environment, sharing of important failure data trumps the secrecy factor. No one wants to see another plant or facility have a fire, an explosion, or an injury, caused by a known defect that could have been prevented by sharing technical knowledge. Therefore, failure data about similar equipment used by others in the same industry is a consideration for the preventive maintenance program at your facility.

It is important to note that this industry data needs to be carefully reviewed before blindly implementing any changes to your program. For example, just because a component caused an event at one facility does not automatically mean that a like component will cause the same event at another facility. You need to know whether it was a design defect or an operating condition that caused the failure. A design defect could be common to any plant using that component. An operating condition where the like component was used in a totally different environment with totally different operating parameters may not necessitate that you change your program. What if valve XYZ was used in a smelter where the failure was caused by temperatures in excess of 1000°F, but you use a similar XYZ valve in a refrigerated area where the temperature never reaches ambient? Industry data functions only as a flag. Further evaluation is required to ascertain its applicability to your facility.

8.1.3.5 Engineering evaluations These are similar to root-cause evaluations, but they are not quite as formal. For facilities that have an engineering staff, it is very common for plant events and equipment failures to trigger an engineering evaluation. Whenever this occurs, a change of some kind to the preventive maintenance is almost always guaranteed. Again, if the component being evaluated has not already been classified as critical or potentially critical, it would most likely be recommended as a commitment component following the engineering evaluation, albeit an internal commitment.

8.1.3.6 Plant design changes Plant design changes and modifications may affect the PM program through the addition of new equipment or through changes to the way the existing equipment operates and functions. It is an obvious input to the RCM analysis to determine the classification for any new components. They range from being critical to being run-to-failure. The RCM process determines whether the new equipment or the new functional design of the existing equipment has an effect on the preventive maintenance program. The COFA Worksheet and the COFA Decision Logic Tree yield the correct resolution.

8.1.3.7 New commitments This input is similar to the commitments that could come about through root-cause evaluations, vendor bulletins, regulatory bulletins, information notices, and engineering evaluations. Depending on your industry, a new commitment may also come about by "edict." It is not an uncommon occurrence for a CEO, vice president, or director to declare a commitment. It would be unlikely, however, if their edict was not already included in your asset reliability criteria, discussed in Chapter 5. If it wasn't included, it now needs to be.

8.1.4 Monitoring and trending

Monitoring and trending is a term that can have very different connotations depending on what it refers to. Monitoring and trending can refer to either individual components or the entire plant. When it refers to individual components, monitoring and trending is performed to detect incipient failures and degradation rates for a component before total failure of the component occurs. It includes performing condition-monitoring PdM activities, analyzing the results of those activities, and comparing them to previous readings to detect trends and rates of equipment degradation. This feedback loop is a very important element for adjusting PM periodicities and for effecting other changes to the PM program such as whether failure of the component can be prevented by different or more frequent planned maintenance.

When monitoring and trending refers to the entire plant, it entails establishing metrics to monitor and trend the overall performance of your plant. Chapter 9 is devoted entirely to monitoring and trending overall plant performance. It is the navigational radar by which you can tell whether your RCM program is performing as it was intended. This data is also extremely important as an input to your living program and provides very insightful feedback on the *effectiveness* of your PM program.

8.1.5 The RCM analysis element

All input elements feed directly into the RCM engine. It is within the RCM analysis element that all decisions regarding changes to the preventive maintenance program are made. All inputs go through the same COFA logic to determine their applicability to the PM program, except, of course, for any new regulatory required commitments that automatically become part of the preventive maintenance program.

The craft feedback, corrective maintenance evaluation, and monitoring and trending elements do not automatically invoke changes to the PM program until they are reviewed and subsequently analyzed in the RCM process. It is important to maintain vigilance about the PMs added to the program or you will find your program escalating with mostly unnecessary work.

The inputs to the RCM engine are dynamic and proactive. The input elements should be continuously evaluated on a real-time basis. This goes beyond the SAE JA1011 document, which recommends only a periodic review. I have positioned the logic for the living program so that you do not have to wait for a periodic review to take action, because doing so places you in a reactive mode by definition.

Your RCM-based preventive maintenance program should remain cutting-edge by incorporating changes as soon as they become evident. This could result in prudent additions to the program, or it could result in deleting PMs from the program. Periodicity changes could also go in either direction. They may be extended, or they may need to be reduced. New PdM technologies could be implemented to eliminate time-directed PMs. The program remains dynamic; hence, it is truly a living program.

8.1.6 Equipment database

The equipment database should always remain current. Each equipment I.D. should be classified according to the COFA logic as to whether it is critical, potentially critical, commitment, economic, or run-to-failure. Some facilities may wish to designate only the classifications of critical, potentially critical, commitment, and economic components and assume all other components are run-to-failure. This is acceptable as long as each component is analyzed and accounted for. Without a specific designation, it cannot be determined whether the undesignated component is an RTF component or an unanalyzed component.

8.1.7 The PM audit

This is an important element to consider. The audit program I am advocating should be periodic. Different facilities and different industries will have their own methodologies for how and when changes are made to the PM program and who makes them. It is highly recommended that one group and one supervisor be responsible and ultimately accountable for *administering* all changes to the program. He or she should be the gatekeeper for all changes. A program where changes can be made by anyone with access to the electronic database or to the index card filing system discussed in Chapter 4 is an invitation to trouble and causes problems. There have been many instances where lack of controls has allowed unauthorized entry to the database to delete tasks, change periodicities, and even add tasks without any justification whatsoever and without anyone knowing about the changes except for the person making them. If this is the type of arrangement your facility has and if you are in a regulated environment, you are in for a hard time. Even if you are not in a regulated environment, this is still not a good practice.

Distinct procedural guidance on who is responsible and accountable for administering all changes to the preventive maintenance program, how the changes are to be documented, and when they occur is an absolute necessity. The program administrator does not have to be the RCM point of contact. The administrator should, however, be the single point of contact for documenting and maintaining the records for all changes.

There is one small challenge for the program administrator. As we learned, the preventive maintenance program is not a maintenance-only program. It is a plantwide preventive maintenance program with various organizations all contributing to performing the various PM activities. Therefore, the recommended PM activities for each component should delineate very clearly which group has responsibility for a given task. For example, if the PM Task Selection Logic Tree credited an operator walk-around as the inspection type of PM for a given component, that responsibility needs to be documented accordingly.

The final PM program that compiles all of the PM tasks for all of the components in the program should clearly identify who is responsible and accountable for performing each task. In many industries, the engineering organization is responsible for PdM activities such as thermography, vibration analysis, and oil analysis. Remember, these are condition-directed PM tasks that are credited with maintaining the reliability of critical equipment. If there is no accountability on the part of whoever performs these tasks and if there is no documentation that these tasks are being performed at the required periodicity, you have a major flaw in your preventive maintenance program.

Oftentimes, the work activities by organizations other than maintenance are not included in a common CMMS system, and any omissions concerning work being accomplished, or when it was accomplished and who accomplished it, escape detection. This is not a good situation. There are many ways to administratively handle this challenge, and it is best left to each facility to determine how to achieve this absolutely imperative requirement.

Just imagine having to explain to senior management how a major unwanted plant event occurred when it was caused by the failure of a component classified as critical. This is especially egregious when the component was supposed to be monitored for incipient failures by a host of recommended PdM tasks to prevent its ultimate failure, but those tasks were never accomplished or were seldom done on time. However, no one recognized this because of administrative loopholes and a lack of accountability in your preventive maintenance program process. Sitting in your vice president's office explaining this event and the reasons for it is not the most desirable position to be in.

In summary, an audit of the PM program should be made periodically to ensure that no unauthorized changes have been made and that all scheduled PM activities, regardless of the responsible organization, have been appropriately accomplished.

8.2 Chapter Summary

Following is a summary of the living program elements:

- A living program is a live program that continues to grow, evolve, change, and adjust.
- It is essential for a living program to have some logic embedded in the process; otherwise, you will oscillate the PM program in a back-and-forth manner by extending periodicities

and then having to reduce them because you extended them too far.
- The craft feedback element allows for the expert opinion and best judgment of the craftsperson or technician performing the work to determine the appropriateness of the task itself. This includes the work scope of the task, the task periodicity, and whether there is a recommendation for some other modification or adjustment to the task.
- There are five craft feedback categories: good (5), above average (4), average (3), below average (2), and poor (1). The feedback is based on the condition of the PM task and not on the overall condition of the equipment.
- The goal is to achieve a category grade of (3), which is average.
- Intelligence has been added to the decision process to account for other influences that can affect the category grade.
- Similar makes and models of equipment used in different systems in different parts of the plant and performing different functions are *not* considered identical, and the feedback data from one cannot be applied to the others.
- The corrective maintenance evaluation element of the living program evaluates corrective maintenance activities to review for the discovery of failure modes that may not have been identified in the original analysis or for new failure modes that were not apparent at the time of the original analysis. It also reviews CM histories to determine whether new PMs should be implemented or changes should be made to the scope of existing PMs.
- A root-cause evaluation is used in most regulated industries such as the airlines and nuclear power, where a component failure that resulted in a significant failure consequence entails a formal root-cause evaluation to not only determine the cause of failure but also ascertain which barriers were insufficient to have allowed the failure to occur.
- A vendor bulletin is issued if a design deficiency or some other type of manufacturing flaw is reported to a vendor and they

are required to take action accordingly. A vendor bulletin should be taken more seriously than the routine vendor manual PM recommendations.

- A regulatory agency generally issues a bulletin—sometimes in the form of an information notice—when something needs immediate attention.
- Many industries share their failure data with each other. Even in a competitive environment, sharing of important failure data overrides the secrecy factor. No one wants another plant or facility to have a fire, an explosion, or an injury caused by a known defect that could have been prevented by sharing technical knowledge. Therefore, failure data about similar equipment used by others in the same industry is a consideration for the preventive maintenance program at your facility.
- In facilities that have an engineering staff, it is very common for plant events and equipment failures to trigger an engineering evaluation. Whenever this occurs, a change of some kind to the preventive maintenance program is almost always guaranteed.
- Plant design changes and modifications may affect the PM program through the addition of new equipment or through changes to the way the existing equipment operates and functions.
- Monitoring and trending plant performance data is the navigational radar by which you can tell whether your RCM program is performing as it was intended.
- All input elements feed directly into the RCM engine. It is within the RCM analysis element that all decisions regarding changes to the preventive maintenance program are made.
- The equipment database should always remain current. Each equipment I.D. should be classified according to the COFA logic, as to whether it is critical, potentially critical, commitment, economic, or run-to-failure.
- It is highly recommended that one group and one supervisor be responsible and ultimately accountable for administering all changes to the preventive maintenance program.

- The recommended PM activities for each component should delineate very clearly which group has responsibility for a given task.
- An audit of the PM program should be made periodically to ensure that no unauthorized changes have been made and that all scheduled PM activities, regardless of the responsible organization, have been appropriately accomplished.

Chapter

9

An RCM Monitoring and Trending Strategy

Chapter 3 briefly explained how to tell whether your plant is reliable. After you have performed a comprehensive RCM analysis and allowed some finite period of time for the program to become embedded, if there is a very clear absence of any serious consequences of failure at your plant, and your workload is utilized primarily for *planned* maintenance activities rather than *unplanned* events, then that is somewhat of an indicator of a healthy reliability program.

Let's go further than settling for "somewhat of an indicator" and delve into a strategy that monitors and measures *aggregate* performance criteria so that you can be proactive in quantitatively assessing your plant reliability. An aggregate philosophy affords you the capability of making quicker decisions instead of reactively responding to make any course corrections if they are needed.

I will show you how to establish vernier navigational metrics so that relatively small adjustments are all that would be necessary to make any needed course corrections to your reliability program. The data required for these metrics can easily be obtained, and if you have a comprehensive CMMS system or internal IT resources, the metrics can be computerized to minimize any labor-intensive efforts. The plant performance calculations and graphs used here can be easily created with a simple Excel spreadsheet.

9.1 What Is Reliability and How Do You Measure It?

Reliability is a commonly used term, but what is it? How do you measure it? Does it encompass just counting the number of plant trips per year to determine whether your plant is reliable or not? Or is it the capacity factor of your plant? Is it the mean time between failure (MTBF) for certain components? As you will see later in this chapter, these generic and rather thin measuring standards, which are the most commonly associated measurements of reliability, can be deceiving and even lull you into a false sense of comfort and security if they are used alone.

Reliability was defined by Nowlan and Heap as "the probability that an item will survive to a specified operating age under specified operating conditions without failure." That is a correct definition in regard to the reliability of a specific *item* based solely on failures. But how do you measure the reliability of the entire entity? What are the precursors to failure that cause concern that the entire plant may not be reliable? Addressing these concerns requires more depth than just looking at reliability as the probability of failure of a given *item*.

Applying Nowlan's definition of reliability to the human body, the reliability of a person could be defined as "the probability that an individual will survive to a specified age under specified living conditions without dying." This is likewise a correct statement. But as we know, there are many subtle precursors to failure that can affect that outcome. For example, how many car accidents was the person involved in and was he or she at fault for most of them? This is an indicator of reckless driving habits and maybe the person won't live to a very old age if that indicator continues at the same level. How many times was the individual overdue for a physical examination? This is an indication of apathy in regard to one's health. How many times was the person cited for serious infractions of the law? This is an indication that the person's behavior may be imprudent and that he or she may not have as many years to live as anticipated. Was the person a smoker or a heavy drinker? Did the person's activities include high-risk sports like skydiving and bungee jumping? These are all valid precursors that provide some insight into the individual's longevity. Can you see how these examples are analogous to precursers at your facility?

Reliability is more than just the probability that an individual item will survive without failure. Likewise, it is more than merely counting gross numbers of failures or the number of lost production days resulting from some type of equipment failure. It is necessary to go beyond Nowlan and Heap's definition and view reliability more as a *measurement of events,* which I define as *the cumulative and integrated rate of unwanted* aggregate *events per unit of time, where the events are not limited to just equipment failures.* By this, I mean that reliability includes a whole host of unwanted events and occurrences that can be measured as a rate of unit operating hours. Reliability represents a broader spectrum of events than just failures, and thus, reliability measurements can offer much more intuitive insight for determining how well your facility is performing.

In many industries, a major fallacy comes into play when comparing one operating entity to another. This is true when comparisons are made without applying a normalizing *reliability performance rate* input to the equation by simply counting the number of failures or the number of power reductions or the number of times a plant incurred an unwanted trip. This fallacy is particularly misleading when you are comparing, for example, the reliability of one plant to that of another. All such comparisons should be normalized by using a rate per x number of unit operating hours.

For example, one plant experiences a certain number of occurrences when it has operated for only 50 percent of the available time compared to the same number of occurrences at another plant that was operational for 100 percent of the comparison period. This is hardly an accurate or fair standard for comparison. Later in this chapter I will introduce you to some typical aggregate performance metrics that I have found to be quite useful for determining how well a facility is being run.

9.2 Monitoring Reliability Is Like Monitoring the Human Body

For years I have observed how inadequate the power generation industry was in regard to monitoring its own effectiveness, as well as its failure to establish an accurate and potent metric to measure reliability performance. It has been a struggle to instill any real intelligence in this issue.

If your facility operates the way the electric utility industry does, your navigation tools for assessing plant performance are probably somewhat crude and rather general. The most commonly referenced measurement factors are metrics such as the number of unplanned shutdowns in the past year and the amount of capacity factor that was lost in the past year. It is not wrong to use these metrics; they are just not good enough.

The metrics I use to monitor the health of a facility are analogous to the metrics for monitoring the health of the human body. The metrics shown in Figure 9.1 can be applied to your entire plant or to each individual system. As noted in Chapter 4, your alphanumeric component database will probably be sortable by some type of system designator. If you have this sort capability, the metrics can be applied on a system basis. If you do not have this sort capability, it is not necessary to create it since the metrics can be applied to the entire facility. As you will see later in this chapter, it is easier to apply the metrics to the entire facility, but there are inherent advantages to monitoring each system individually. Once again, the choice is yours.

9.3 Caution: Avoid Analysis Paralysis Performance Monitoring

I have mentioned several times that any program, whether it is an RCM program or a monitoring and trending program, can be made extremely difficult and complex or very robust and simple. I prefer the latter. A powerful example can be found in the nuclear power industry. As sophisticated as this industry may seem to be, the navigation tools it used for many years to track its own performance were based on two primary metrics: (1) how many plant trips occurred? and (2) what was the plant capacity

The all-encompassing (and inadequate) monitoring criteria:

For the facility:

- Plant trips
- Capacity factor

For the human body:

- Temperature
- Blood pressure and pulse

A more detailed performance monitoring strategy would include criteria such as:

For the facility:

- Plant trips
- Capacity factor

   Plus

- Unplanned operator actions
- Unplanned power reductions
- Production delays
- Enforcement actions
- Litigation occurrences
- Citations or violations
- Root-cause evaluations
- Injuries
- CMs written
- Overdue CM backlog
- Overdue PM backlog
- Other

For the human body:

- Temperature
- Blood pressure and pulse

   Plus

- The heart
- An EKG of the heart
- The lungs
- The kidneys
- The arteries
- The thyroid
- The liver
- The spleen
- Blood chemistry
- Eyesight
- Hearing
- Other

Figure 9.1 Monitoring the health of a facility is analogous to monitoring the health of the human body.

factor? There were several other metrics, but these were the primary ones.

When some nuclear plants decided they needed additional metrics, they did what most facilities would do that have an abundance of engineering personnel available. The simple became the complex, and the monitoring program became an empire unto itself. What had been a one-page metric became a metric "book" consisting of several hundred pages, which for the most part was unusable and impractical. The metrics consisted of innumerable subjective opinions, attitudes, and beliefs, accompanied by extensive excuses to justify poor performance.

Performance metrics are intended to provide senior management with a snapshot of how reliably their asset is performing. Vice presidents do not have two weeks to read a "book" filled with opinions to ferret out that answer. The metrics should be simple, comprehensive, and objective, and they should provide an aggregate snapshot of reliability.

9.4 The Aggregate Metrics

Think how comprehensive your medical examination would be if the only monitoring that took place was to record your temperature, blood pressure, and pulse. That would hardly constitute a thorough physical examination to determine your level of health and the performance of your body. What about your heart rhythm, kidney function, lung capacity, blood chemistry, bone density, and so on?

As unlikely as it may seem, the extent of monitoring that many industries use to determine the health and reliability of their facilities is analogous to the limited monitoring of your physical well-being that would include only your temperature, blood pressure, and pulse. That is way too Spartan in today's complex environment to effectively evaluate the performance and reliability of a facility.

The metrics I used to create the monitoring and trending strategy include all of the possible events and performance data that can have an impact on plant reliability. Unplanned trips and capacity factors are included in the metrics, but they are not the only considerations. The events I refer to are those caused by equipment problems or an inadequate preventive maintenance program; they are not events caused by the weather, an act of God, or human intervention. I have attempted to integrate all of the possible dynamics that can occur, with a weighting factor incorporated for each one depending on the relative importance of the event or occurrence.

The monitoring and trending strategy is simple to understand and simple to use, and the results can be computed and graphed automatically even within a basic Excel spreadsheet. The ultimate plant performance rate is made to be as objective as possible. As long as the weighting factor remains constant for each event type, the calculated six-month moving average of your aggregate plant performance rate will maintain its degree of consistency and objectivity so that it becomes a true trending comparison of your system and/or plant rate of performance from one time interval to the next, whether it be week to week, month to month, or year to year.

Furthermore, reliability should be measured per a unit of time. It is more appropriately measured as a rate per 1000 unit operating hours, for example. Therefore, the aggregate metrics I have compiled are computed for a given rate of time. I do not believe that performance should be based on a static gross number of occurrences. Reliability can thus be measured in quantitative terms rather than based on opinions or loose interpretations and predictions by those responsible for assembling and managing the monitoring program. If objectivity is missing from the monitoring and trending strategy, it becomes too easy to manipulate and skew the results.

Let's look at each one of the metrics.

9.4.1 Unplanned plant or facility trips

Unplanned trips are a metric weighted with a relatively heavy factor. This is an event that should not be occurring with any regularity. It results in a total loss of production, generating capability, or mission objective, and in many instances it triggers the actuation of emergency systems. In an automobile manufacturing plant, how many times did the assembly line shut down? How many times did a cruise ship require a tugboat escort back to harbor? How many times did an electrical generating unit shut down? How many satellite launches were aborted? How many flights were cancelled? These are all considered plant trips.

An excessive number of these events per year is an indication that your facility is not operating as reliably as possible, and improvement in your preventive maintenance program or possibly some design changes may be needed if the PM program alone is unable to eliminate these occurrences.

9.4.2 Capacity factor

How much less than 100 percent of the available time did your plant operate, not counting planned outage time and other planned downtime? Obviously plant trips, power reductions, and production delays are functions of this metric. This is an acceptable metric, but all too often it is used in isolation, and when used by itself, it does not provide a realistic indication of overall plant performance. Later in this chapter you will see how this metric by itself can be quite misleading.

9.4.3 Unplanned operator actions

This is an indicator of the number of times plant operators have been required to take some type of unplanned action to circumvent an unwanted occurrence or take other compensatory actions as the result of a component failure or some other deficient plant condition. In many industries, a logbook is maintained to capture these unplanned operator events.

This is a subtle metric that is often overlooked but can be very enlightening for revealing precursors to unwanted plant consequences and underlying equipment problems. For example, does each day present a myriad of challenges to the plant operators, where they are continually trying to avert some major unwanted event, or does the facility run rather smoothly on a daily basis? There will be normally expected operator evolutions as a course of routine operation of the plant, and those events are not considered in this metric. Only those unexpected occurrences that require some type of operator action that is totally out of the ordinary routine are part of this metric.

Such an occurrence might be an immediate action required on the part of the operator to avert a serious plant consequence, or it might be an operator work-around. Operator work-arounds are not considered normal practice, and they become a distraction and diversion from normal operator duties. For example, a work-around could involve the manual operation of a component or system that normally operates on auto. Continually having to establish alternate operating parameters or other actions to compensate for plant deficiencies can negatively influence the reaction time in the event of a real emergency. This is an important indication that a preventive maintenance program is in need of more attention.

9.4.4 Unplanned power reductions

How many times has your facility had to downpower in order to gain access to and clearance for the repair or replacement of a failed component? Many components cannot be worked on under full-power conditions. Another consideration is how many times a component has failed that required the immediate downpower of the plant to prevent it from tripping offline. A plant that constantly has to downpower for unplanned reasons is obviously not operating at its optimum reliability level. Power reductions can last for minutes, hours, or even days and weeks. The time duration threshold for including this metric is based on the asset reliability criteria and the qualifying condition of time you selected. A *pattern* of unplanned power reductions, for whatever reason, that are in excess of several hours per occurrence is not an indication of a reliable plant.

9.4.5 Production delays

Does your facility constantly experience nuisance production delays even for short durations of time? These can be any one of a host of different types of delays, such as assembly line delays, departure delays, launch delays, or manufacturing delays. Even though each one of these may be of a short duration, there is something telling about having an abundance of them. An excessive number of these delays indicates a preventive maintenance deficiency.

9.4.6 Enforcement actions

If your plant is subject to enforcement actions as a result of plant performance issues, regardless of the reasons for them and regardless of how well you believe your plant is operating, you should consider them a wake-up call telling you that all is not well with your preventive maintenance program. Enforcement actions may occur as the result of an accumulation of insignificant infractions that, when looked at in their totality, signify to the regulators that something is amiss. In regulated industries, enforcement actions are a decidedly undesirable occurrence since they can bring down the wrath of the regulators upon you and cause you to be subject to more oversight and scrutiny than you are accustomed to.

The regulators are continually getting smarter. Regulated facilities are subject to periodic routine inspections, and enforcement actions are quite common when the regulators perceive an undercurrent of problems. Even though a less visionary management might try to sugarcoat the real issues by arguing that everything is okay because the plant is running at 99 percent capacity, it is only a matter of time before those underlying problems not only result in a diminution of your capacity factor but create the real potential for a major unwanted event to occur.

Usually, enforcement actions are the result of a series of small events that are really precursors to larger unwanted events. Even though your capacity factor may be high and you may not have had any plant trips (remember, this is analogous to looking only at your blood pressure and pulse rate), that does not inoculate your facility against enforcement actions. This is another subtle metric that is usually dismissed if the plant is perceived to be running at maximum output.

9.4.7 Litigation occurrences

These events are most undesirable, and for regulated industries they go hand in hand with enforcement actions. These are litigation occurrences caused by equipment failures, not litigation as a result of patent infringement, for example, or any other litigation outside the boundary of cause by the preventive maintenance program. Chapter 1 mentioned that a famous theme park was in the headlines of the major newspapers for litigation that resulted from injuries caused by mechanical malfunction. The malfunction itself may have been considered insignificant by the theme park corporation and downplayed as not being a serious mechanical flaw, but the fallout was on the front page. Any future "cut fingers" at that theme park caused by mechanical failure would likely make at least local headlines. This is another precursor that presents an obvious opportunity to put additional emphasis on the preventive maintenance program.

9.4.8 Citations and violations

Citations and violations constitute another subtle metric that often passes undetected as a precursor that something is wrong. You do not have to experience a host of equipment failures to accumulate a number of citations and violations. As with enforcement actions, even though your capacity factor might be high and you might not have had any plant trips, that does not immunize your facility against citations and violations. Citations and violations may occur as the result of an accumulation of small, inconsequential infractions when looked at individually, but when looked at in their entirety, they present a very clear picture that deficiencies exist in your preventive maintenance program.

Accruing citations, violations, and enforcement actions can be an indicator that your management does not take preventive maintenance seriously enough. As Chapter 3 mentioned, it is not uncommon for management to neatly package or whitewash an accumulation of small events as individual isolated incidents rather than viewing them cumulatively as the possible existence of a much broader deficiency in your plant's reliability.

9.4.9 Root-cause evaluations

Root-cause evaluations are usually reserved for the more significant plant events. Whenever you find that root-cause evaluations are becoming more common in your facility, it is time to take a hard look at your maintenance practices. A reliable plant should not be stacking up a multitude of these evaluations, as it is an indicator of underlying problems with your maintenance program. Once again, it is not uncommon—and in fact it is quite common—that the plant capacity factor is high and the number of plant trips zero, but the metric precursors nevertheless reveal a poorly run plant.

9.4.10 Injuries

This metric is a very good indicator that a facility is either run well or not run so well. Injuries can occur either to employees or to members of the public in the case of facilities that cater to the public such as theme parks, cruise ships, and airlines. Personnel injuries of any kind must be avoided as a foremost priority. It is an accepted fact that some injuries are caused by irresponsible behavior on the part of the employee or the public. For example, an employee stands on the topmost rung of a ladder when he or she knows not to; thus, a subsequent fall would be due to the employee's carelessness. Similarly, if a member of the public attempts to perform some unauthorized acrobatic stunt while on a ride at a theme park and is injured in the process, this injury would also be the result of personal carelessness.

The injuries that are the indicators of a poor reliability program occur when the employee is following all of the proper procedures and still incurs an injury. The incidence of any injury, even if it is not considered serious, is an indication of the potential for more serious injury consequences. Any injury that was not the result of personal carelessness must be taken seriously.

Even if carelessness was the cause of the injury in question, you must consider these issues: Did the facility take every measure to ensure that appropriate warnings were in place to alert employees and/or the public to such risks and take measures to safeguard them? Were these warnings posted in public places? Were employee training programs adequate to handle the situation?

9.4.11 Rate of written CMs

A certain number of CMs each day are a routine expectation. This is a difficult metric to define. I include it because it does provide some insight into the rate at which CMs are being written—although CMs are not necessarily a negative attribute.

In fact, many plants consider that when a deficiency is found by a PdM activity, the deficiency then becomes a CM. This is an acceptable practice.

Usually CMs are prioritized according to their importance. For example, priority 1 CMs usually involve deficiencies that must be corrected immediately such as steam leaks, oil leaks, or personnel hazards. Priority 2 CMs require attention usually within one week. Priority 3 CMs commonly include corrective maintenance that can wait until the 12-week planning cycle recurs to be scheduled for work. The weighting factors are slightly different for each priority of CM.

9.4.12 Overdue CM backlog

While the number of CMs written is not necessarily a negative attribute, the inability to manage them once they have been created certainly is. The inability to manage CMs becomes evident when reviewing the overdue CM backlog. CMs that start to stack up without being planned and completed when they are scheduled is an excellent metric for judging how capable your facility's resources are to handle the work. If corrective maintenance work cannot be accomplished when it is scheduled, it is only a matter of time before some unwanted plant event occurs. As we learned in Chapter 3, corrective maintenance is an integral part of the bigger picture of preventive maintenance, and addressing CMs ultimately becomes as important as addressing PMs.

A growing backlog of overdue CMs should be a warning sign of imminent danger. If your facility cannot handle the CM workload being generated, playing ostrich is not an option. You will need to ascertain whether your organization is inefficient in its ability to plan, schedule, and accomplish routinely generated work or whether your workforce is understaffed and cannot cope with the workload. My experience indicates that most often it is the former rather than the latter.

9.4.13 Overdue PM backlog

This is a very important metric. An efficiently operating plant will have very few, if any, overdue PMs. A plant that is in trouble will have a whole host of overdue PMs. It does not take a lot of insight to realize the jeopardy your facility will be placed in if your organization cannot cope with the need to complete PMs when they are due.

If your facility cannot handle the PM workload, your organization is either inefficient or understaffed. As before, it is most often the former rather than the latter that causes an overdue PM backlog.

In some regulated industries, a backlog of overdue PMs is sufficient reason for receiving a citation or a notice of violation from the regulators. This metric can be a judgment call by a near-sighted management who tend to argue that all is well as long as the plant is operating at or near full capacity.

Like a growing backlog of overdue CMs, any backlog of overdue PMs should also be a warning sign that danger is ahead. In some industries, there is a grace period for PMs that normally allows an extension of up to 25 percent of the PM period to accomplish the task. For example, if the periodicity is once per year, the allowable extension of 25 percent would be three months. Therefore, the PM would not be considered delinquent until it is overdue for one year plus three months.

The 25 percent grace period is an allowable and acceptable time frame extension to accommodate unplanned scheduling problems. Good plants use this allowable extension sparingly, as it was intended. Other plants automatically use the 25 percent allowable extension as part of the normal planning process and then find that they have delinquent PMs to deal with. Depending on your industry, you may or may not be familiar with the grace period concept. If you are not, it doesn't matter, because you are already scheduling your work according to the specified periodicity.

If your industry does use a grace period for PMs, note that the metric I use for PMs applies to any PM that is overdue according to its actual periodicity. It is not a metric for delinquent PMs that are 25 percent beyond the point of being overdue. I deliberately defined it that way to circumvent any imprudent manipulation intended to avoid having to complete a PM on its due date.

9.5 Weighting Factors

By now you should have a good understanding of why the aggregate metrics I have presented here are much more insightful when you are determining the health of your facility than merely keeping a count of the number of plant trips and the capacity factor. Those metrics alone do not present a complete picture.

The weighting factors for the metrics should be developed by a consensus of the engineering, maintenance, and operations representatives and should reflect the relative importance of each event. Some event metrics will specify a certain number of expected occurrences such as the number of CMs written or the number of CMs in the backlog. Therefore, the organization representatives should also specify the expected number of event occurrences per month, per quarter, or per year. This establishes management's expectations for performance comparisons from one time interval to the next interval. Obviously, the number of expected occurrences for some events should be zero—such as for the number of plant trips, citations and violations, or injuries. I strongly recommend that the weighting factors and performance expectations be formalized in a document signed by the managers of the three participating organizations.

The weighting factors are arbitrary in that they serve only to compare relative importance. Your weighting factors should be what you and your management deem prudent. Refer to Figure 9.2 for some typical examples of weighting factors and expected numbers of occurrences.

9.6 Performance Calculations

The performance calculations are based on a rate of time. In the example in Figure 9.2, the expected number of occurrences and the actual number of occurrences are per quarter for the entire plant—although it could be per week, per month, per year, or whatever you choose. You could also set it up on a system basis. The metrics in Figure 9.2 are per occurrence. For example, the weighting factor of 2.0 for a plant trip is multiplied by the number of occurrences per quarter.

As noted, some of the metrics will indicate expected numbers of occurrences per quarter as shown in Figure 9.2. Each priority 1 CM that was overdue for between one and three months has a weighting factor of 0.010 per CM. The weighting factors can be whatever you wish them to be as long as they represent the rel-

Metric	Weighting Factor	Expected Number of Occurrences in a Quarter	Actual Number of Occurrences in a Quarter
Plant trip	2.00	0	1
Capacity factor:			
■ For each 1% below 100%	0.50	1.5	2.0
Unplanned operator action	0.50	1	2
Unplanned power reduction:			
■ Greater than 25% power reduction	0.75	1	1
■ Greater than 15%, less than 25%	0.50	1	2
■ Less than 15%	0.25	2	2
Production delay	0.15	2	4
Enforcement action	2.50	0	0
Litigation occurrence	1.50	0	0
Citation or violation	2.00	0	1
Injury (serious)	2.00	0	1
CMs written			
■ Priority 1	0.0015	150	300
■ Lower than priority 1	0.0010	250	200

Figure 9.2 Performance monitoring and trending.

Metric	Weighting Factor	Expected Number of Occurrences in a Quarter	Actual Number of Occurrences in a Quarter
Overdue CM backlog			
Priority 1:			
■ Overdue 1–3 months	0.010	15	25
■ Overdue 3–6 months	0.015	10	15
■ Overdue more than 6 months	0.020	5	15
Lower than priority 1:			
■ Overdue 1–3 months	0.005	25	35
■ Overdue 3–6 months	0.007	20	25
■ Overdue more than 6 months	0.010	10	10
Overdue PM backlog:			
■ 25–50% overdue	0.25	2	5
■ 51–100% overdue	0.50	1	2
■ >100% overdue	0.75	0	0
Other, depending on the type of industry or facility.			

Figure 9.2 Performance monitoring and trending. *(continued)*

ative importance of each metric. As you can see, an enforcement action is weighted more heavily than a plant trip or a citation or violation. A plant trip or citation or violation is weighted more than a power reduction. The overdue CM backlog is weighted more heavily than the rate of CMs written.

The performance monitoring and trending calculations can be applied to the entire plant or to each individual system. Using the individual system basis provides a more diverse look at the reliability picture and breaks it down to a lower level. If you have the capability to identify the individual components that triggered each of the metrics, by system, then you will be able to take a more detailed snapshot of performance. If this capability does not exist, your performance monitoring can remain at the plant level to provide an overall snapshot of performance.

If your plant was not operating due to a planned outage or even unplanned downtime, those hours are not included. The reason for this is that if you had only one unplanned operator action in the quarter, for example, and your plant was not operating for two months of the quarter, the effective performance rate of that one occurrence would not be the same as if the plant had been operating for the entire quarter. Looking at it another way, one could say that the plant had only one occurrence in the entire quarter, even though it was in operation for only four weeks out of the entire quarter. Calculations that do not take the element of time into consideration can result in an inaccurate standard of measurement.

The performance rate calculations are shown in Figure 9.3. There is an *expected performance rate* (EPR) and an *actual performance rate* (APR). The EPR is the rate per 1000 plant operating hours based on 24 hours/day × 90 days/quarter = 2160 hours in the quarter. The EPR would be the sum of all of the expected numbers of occurrences multiplied by their respective weighting factors times 1000 and divided by 2160. As shown in Figure 9.3, the *sum* of all of the EPR metrics multiplied by the respective weighting factors is 5.54. Therefore, the expected performance rate is 5.54 × 1000/2160 = 2.57.

The APR would be the sum of all of the actual numbers of occurrences multiplied by their respective weighting factors times 1000 but divided by the actual number of operating hours

> A. From Figure 9.2, the *sum* of all of the *expected* number of occurrences multiplied by their respective weighting factors = 5.54
>
> Therefore, the *expected performance rate* (EPR) per 1000 unit hours =
>
> 5.54 × 1000 / 2160 = 2.57
>
> B. Also from Figure 9.2, the *sum* of all of the *actual* number of occurrences multiplied by their respective weighting factors = 14.98
>
> Therefore, the *actual performance rate* (APR) per 1000 unit hours =
>
> 14.98 × 1000 / 2000 = 7.49
>
> Notes: 1. There is a total of 2160 available hours in a quarter.
> 2. There was a total of 2000 actual operating hours in the current quarter.

Figure 9.3 Performance monitoring calculations.

in the quarter, which was 2000 hours in this example. As shown in Figure 9.3, the *sum* of all of the APR metrics multiplied by the respective weighting factors is 14.98. Therefore, the actual performance rate is 14.98 × 1000/2000 = 7.49.

9.7 Performance Graph

In Figure 9.4, I have shown a typical performance trend line for the entire plant. This could also reflect the trend for each system. Let's look at the attributes of this chart. The plant EPR was calculated to be 2.57. The quarterly calculations are trended on a six-month moving average to remove any skewed oscillations and perturbations. Remember, it is the relative comparison of quarter to quarter that is important to detect any negative trends or to acknowledge any positive trends. Once you have developed the database for storing the metric numbers, the data can be very simply calculated and graphed automatically on an Excel spreadsheet or any other software program. You may even wish to trend your data on a monthly basis with a three-month moving average.

Chapter Nine

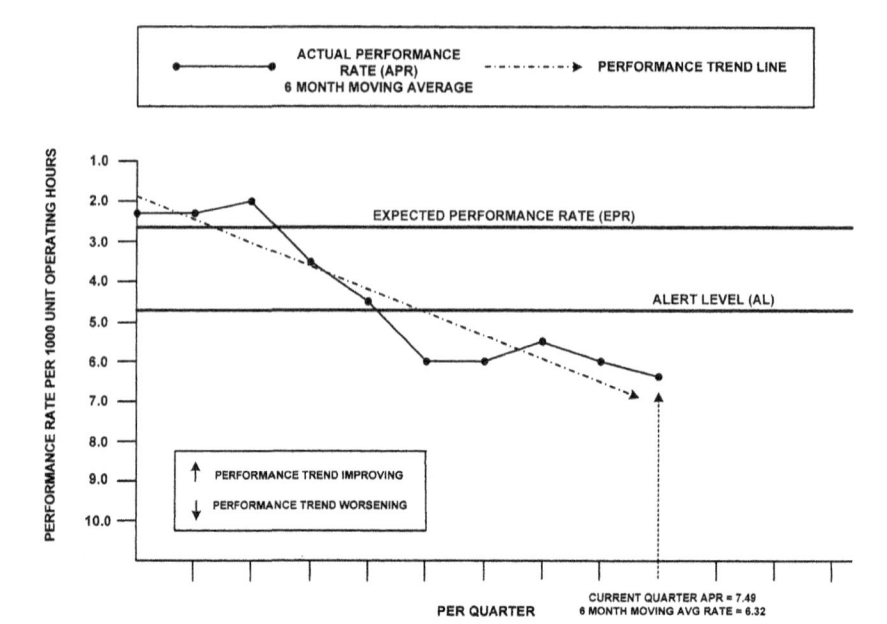

Figure 9.4 A typical performance trend.

I have also included an *alert level* (AL) below the expected performance rate to raise a flag once your moving average trend line crosses this level. The AL can be set at whatever level you choose. It is intended to provide a red flag indicating that your performance has passed a threshold into an unacceptable zone. You may even choose to set your AL the same as your expected performance rate so that any deviation from the EPR is deemed unacceptable.

In this example, the plant APR for the current quarter was calculated to be 7.49, and the six-month moving average is 6.32. As you can see, the trend for this specific example plant is not favorable. In fact, it indicates a plant that is in trouble. A performance trend line as shown in Figure 9.4 is indicative of a worsening trend.

Each month or each quarter, this performance trend can be reviewed by senior management to provide a snapshot of how well the plant is performing relative to either a positive or a negative trend. The metrics are totally objective, as are the weighting factors, as long as they remain constant. The only subjectivity that enters the equation is the determination of the number of expected occurrences.

What is most interesting to note is that a worsening performance trend, as depicted in Figure 9.4, is not atypical of a plant running at a 99 percent capacity level. It is that false sense of comfort when operating at the 99 percent capacity level that can camouflage the aggregate metrics that clearly reveal a deteriorating facility. The deteriorating condition would otherwise go unchecked until a major unwanted event or a series of events occurs. It has been my experience in the airline industry and in the nuclear industry that many airlines and nuclear facilities that once boasted a 99 percent capacity factor without any idea of their underlying problems eventually ended up on a regulator's watch list within three to five years, if not sooner. This has happened all too often.

A regulated facility that is heading toward a position on the infamous watch list or any facility that ultimately finds itself in deep trouble in regard to its maintenance practices usually does not get there overnight. It is the cumulative effect of a series of preventive maintenance issues that leads to a slow but steady decline in performance until it is too late to correct the situation without a drastic upheaval, possibly even leading to a change in the plant management as a corrective action. This steady decline usually takes place under the radar. That is why a robust monitoring and trending program that includes an aggregate of metrics is so important in preventing that unwanted slide into a reliability ravine.

Depending on your specific industry or type of facility, you may wish to include other metrics in your trending program—for example, departure delays and cancellations for an airline, or the number of lost barrels of oil for an oil drilling enterprise. You may also wish to use different weighting factors.

Remember, *the objective of the monitoring and trending program is to discern any relative change in performance so that corrective measures can be immediately implemented.*

9.8 Performance Graph by System

Figure 9.5 shows what individual system performance trends would look like. These are snapshots per month or per quarter computed for each system, similar to the plant performance trend shown in Figure 9.4. The methodology and philosophy are

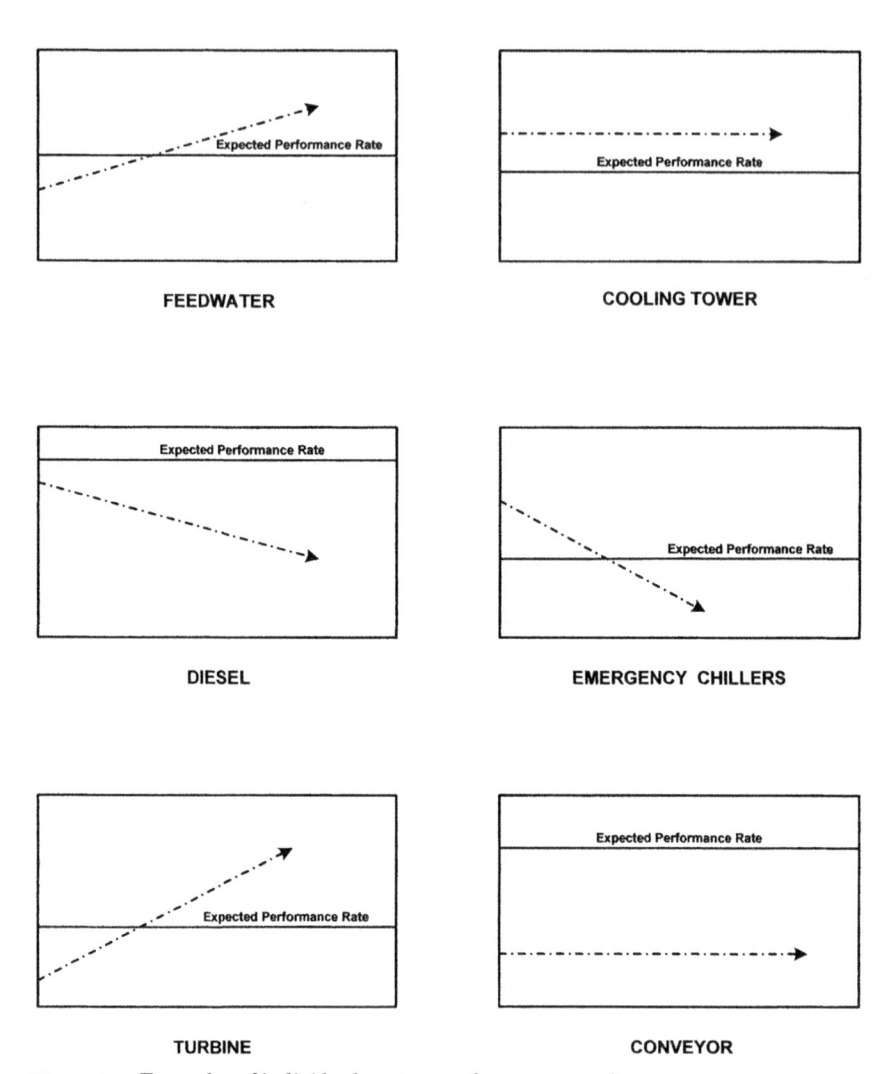

Figure 9.5 Examples of individual system performance trends.

the same. The only differences are that for an individual system performance trend, you need to know the expected number of occurrences by system and identify the components by system so that the metrics are triggered by the components within that respective system. For example, the CM and PM data would be applicable to individual components, and they would need to be identified by system. Similarly, if a component failure resulted in a plant trip or power reduction, you would need to know what system that component belonged to.

If you elect to trend your performance by system, your management will quickly be able to determine at a glance which systems are in need of extra support and which ones appear to be operating reliably. It can afford your organization the opportunity to reallocate resources accordingly, as well as pinpoint incipient worsening trends by system to avert problems before they occur.

9.9 A Final Caution

Figure 9.6 raises this question: Can your facility be operating at a 99 percent capacity factor with no plant trips but claim ownership of the same system trends shown in Figure 9.6? The answer is: *absolutely!* Avoiding this scenario is what an *aggregate monitoring and trending* program is all about.

9.10 Benchmarking

I encourage the proper use of benchmarking initiatives. Benchmarking is a technique that you may already be familiar with. It consists of sending a person or a team to another facility to find out how they are conducting business and see whether there is something of value to learn from the experience of others. This can be worthwhile, but I caution you about what I refer to as "incestuous" benchmarking, a practice that I have witnessed too often.

This occurs when plant A sends a team to benchmark plant B to determine their best practices. What plant A does not know is that plant B just visited plant C to find out how they are conducting business. You guessed it: plant C had just returned from a benchmarking trip to plant A, and plant A has therefore just benchmarked itself! Sadly, this process is symptomatic of an unenlightened management team.

I recall a rather humorous experience in which a senior manager mentioned that he had just found out that facility X had a really good practice that they had touted to him. After he finished explaining the great things that facility X was doing, I diplomatically replied that their engineering manager had just visited and the practices they were touting came from us!

240 Chapter Nine

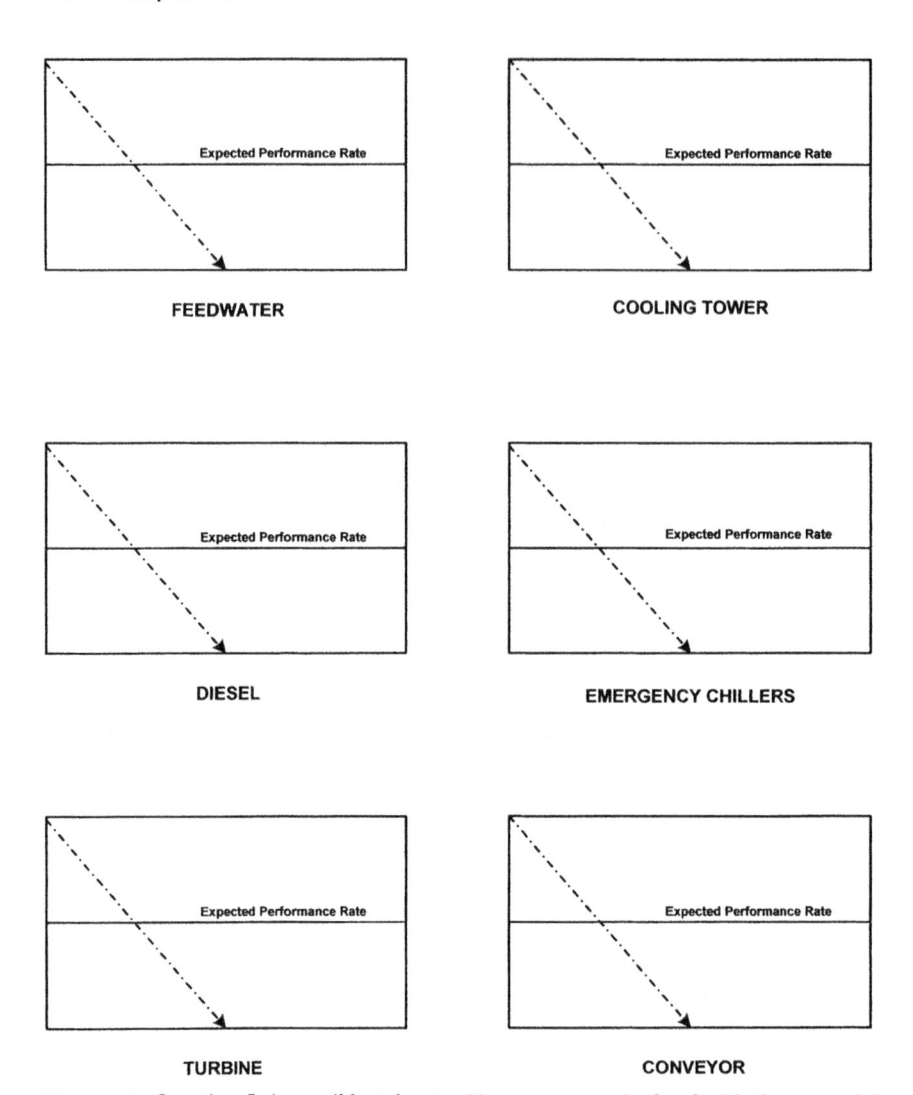

Figure 9.6 *Question:* Is it possible to be at a 99 percent capacity level with these trends? *Answer:* absolutely!

Once you have implemented a simple but robust premier RCM program and an aggregate monitoring and trending program, you most likely will find that others will be knocking on your door for an invitation to learn how you established these practices.

9.11 More About Expected Performance Rates

Some facilities elect to absorb one plant trip per year. Therefore, rather than zero, the EPR for plant trips per quarter would be $2.00 \times 0.25 = 0.50$. A plant may also choose to accept eight power reductions per year with each one being greater than 25 percent. Therefore, rather than zero, the EPR increment for power reductions per quarter would be $0.75 \times 2 = 1.50$.

It is the mind-set of higher management that determines the negative influences they are willing to accept as a cost of doing business. Some management may settle for as many as four newspaper headlines per year proclaiming negative publicity about major unwanted events that have occurred in their facility. Some may accept that 10 plant trips per year are acceptable, or that a certain number of serious injuries are acceptable. When a facility is too accommodating and accepts a lax EPR, that facility is not a premier example of reliability. In fact, if such a management mind-set exists, embarking on an RCM program may prove to be a wasted effort.

9.12 Avoid Reliability Complacency

Facilities that continuously raise the bar each year, striving for even better performance, and facilities that embrace a very strict EPR are better plants that others should try to emulate. RCM is a perfect fit for their objectives. When the performance bar is constantly being raised, maintaining the status quo is not an option unless you want to fall behind the reliability leaders.

Unfortunately, and all too often, a certain smugness about being a good performer pervades senior management meetings. That is the point at which they typically believe they have achieved reliability nirvana and begin to allow their facility to slip backward because they cease to be vigilant about their preventive maintenance program. In my experience, this is a common pitfall and should most definitely be avoided. I have seen extremely well-run plants with stellar performance slide into oblivion because of management complacency.

Plant reliability was not intended to be put on autopilot. It requires constant attention and involvement. An absence of undesirable plant consequences for a mature plant means that someone has been doing something right. It is usually at these peak performance times that management attempts to tweak the system and begins an experimentation process to further reduce costs by cutting back on what they perceive as excess resources because the plant is operating so well. This road to reliability despair is more apt to occur when a new management team arrives, whose first order of business is to improve the bottom line.

Alternatively, I have seen plant managers who were too deeply entrenched in their positions and were highly reluctant to introduce any changes for fear that they would be criticized for not implementing them sooner. Instead of doing what the situation demands, self-preservation and personal ego considerations often override prudent and sensible managerial decisions.

When you finally achieve a respectable reliability performance profile for your plant, the maintenance program can become a target-rich opportunity for corporate bean counters. To their way of thinking, since everything is running so well, there obviously must be room for cutting costs—which they proceed to do without any regard to the consequences. This is a major mistake—albeit one that occurs with too much regularity. Let's look at how this is depicted graphically.

9.13 How to Maintain Your Reliability Performance

As you ramp up your RCM effort, reliability is your number one objective. As Chapter 1 notes, RCM is a reliability program, not a PM reduction program. If you started with a Spartan preventive maintenance program, you will no doubt be adding work that was not being done but should have been done, and your staffing needs will most likely increase.

Suppose yours is one of the facilities that performs too much unnecessary work—for example, there are too many intrusive PMs for periodic overhauls and replacements rather than utilizing more PdM activities. A sound RCM program affords you the opportunity to become more efficient. This does not mean the immediate layoff of 15 percent of the workforce if you eliminated 15 percent of the PMs. After allowing some time for your RCM program to integrate itself into your work procedures, the situation in your facility will most likely resemble the graphic representation in Figure 9.7, which shows a relationship between maintenance worker-hours (WHs) and time whereby the maintenance WHs required are the sum of the workloads for accomplishing both PMs and CMs.

If you have found that your RCM effort has identified a larger population of equipment than you had before that can either be categorized as run-to-failure or be subjected to PdM tasks as opposed to intrusive overhauls, your PM worker-hour requirements will decrease. Why? One reason is that if you are performing PMs on equipment that is found to be acceptable to RTF, there will obviously be fewer PM worker-hours needed. Additionally, if you have exchanged intrusive tasks for PdM tasks, your PM staffing requirements will also be reduced. But wait, that is not the end of the story. Your *CM* staffing requirements will increase! The good news is that they will increase at a rate that is less than the rate at which PM staffing will decrease

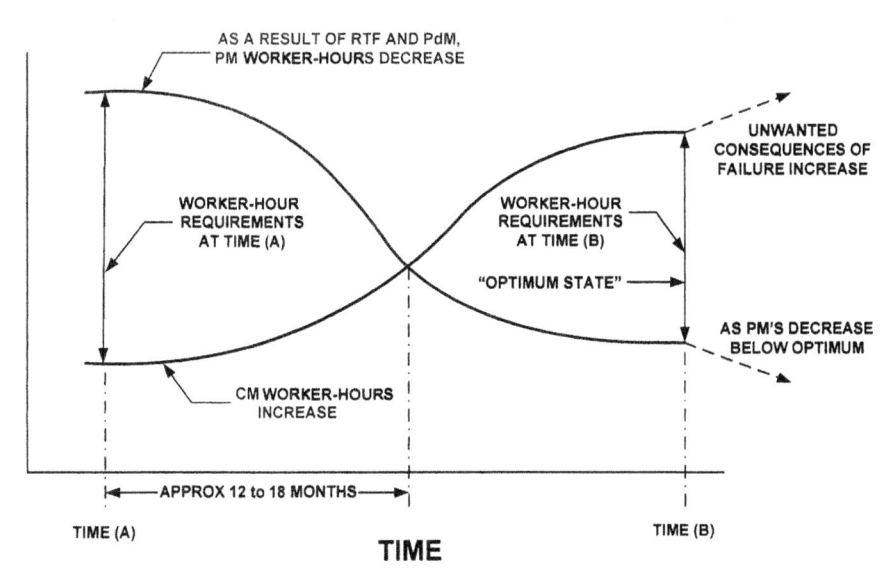

Figure 9.7 Maintenance worker-hours (PMs + CMs).

until a certain time interval, which from experience can be estimated at 12 to 18 months.

At some point the PMs and CMs will reach a state of equilibrium. This is shown in Figure 9.7 as the "optimum state." Here are some considerations relative to RTF, PMs, CMs, PdMs, and staffing:

- With a greater understanding of the RTF concept and how RTF can become an integral part of your program, RTF-generated CMs will be mostly "emergent" work rather than "planned" work such as would be the expected outcome from PMs and PdM tasks.
- With PdM activities, CMs can be planned and scheduled accordingly because of the advance warning of an impending failure.

Again, look at Figure 9.7. With any newly implemented RTF philosophy, PMs will decrease rather rapidly; thus, PM worker-hours will also decrease rapidly and the number of CMs will begin a rapid ascent for approximately 12 to 18 months. Worker-hours will then begin to stabilize. The overall resultant maintenance worker-hour requirement (PMs + CMs) at time (B) will probably be less than they were at time (A) because of the efficiency of working on equipment when work is needed rather than performing unnecessary PM work when it is not needed. A component will still need to be overhauled at some point, but the overhaul won't be done until the component needs it, and that is where much of the maintenance efficiency takes place.

This assessment is further supported universally by the "bathtub" curve, whereby only 11 percent of all equipment shows a definitive wear-out pattern—in other words, 89 percent of all equipment otherwise fails randomly. This underscores the importance of performing PdM accordingly, as discussed in Chapter 6.

Perhaps one of the most significant issues that industry faces is understanding the impact that PdM and RTF have "downstream." More to the point is that in the future much of your work will no longer be driven by PMs but instead will be driven primarily by emergent corrective maintenance. This is not a bad

An RCM Monitoring and Trending Strategy 245

thing. In fact, as shown in Figure 9.7, it is a more efficient way of performing maintenance.

The optimum state of equilibrium continuously morphs into relative states of equilibrium as your plant ages, as new failure modes become evident, as plant modifications occur, and as newer PdM activities become available and get incorporated into your program. The optimum state may shift relatively, but the same relative equilibrium prevails. The optimum state is unique for each facility, but in relative measures it is similar. This is like a discrete signature for each plant. It is the functional balance of the unique design of your plant, its equipment, and its operating practices. Your goal is not to have zero CMs; rather, it is to seek the optimum point of equilibrium between PMs and CMs that is specifically applicable to your facility—in the same way that your signature or fingerprint is unique to you.

Figure 9.8 visually depicts how delicate the balance is in achieving the optimum state of equilibrium. The preventive maintenance strategy incorporates PMs, CMs, PdMs, RTF, and even design changes that work in harmony to balance out any unwanted consequences of failure or other unplanned events. The size of the "seesaw," the weight of the given "loads," and the location of the fulcrum point are unique for each industry and for each facility within that industry.

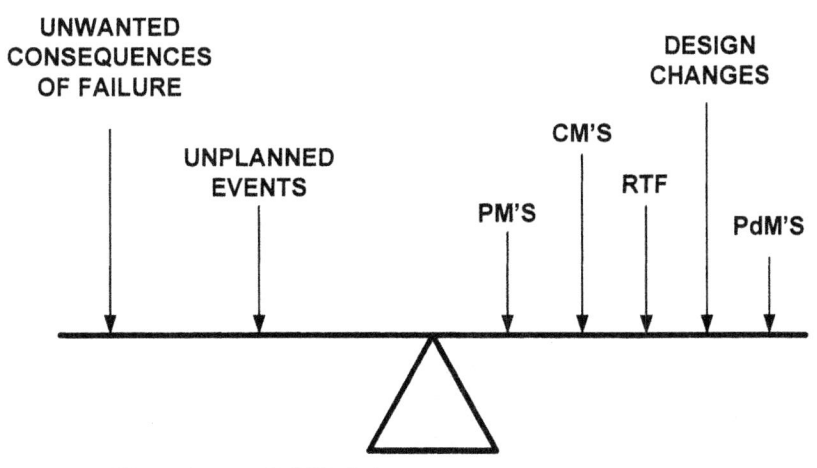

Figure 9.8 The optimum reliability balance.

Indiscriminately tweaking the preventive maintenance program below the optimum level without regard to the implications of doing so results in a diminution of your overall reliability levels. Indiscriminately tweaking a premier reliability program is not unlike adding an extra measure of salt or sugar to a proven recipe. It can totally upset the balance of that recipe, making a connoisseur's delight into something inedible. As shown in Figure 9.7, once PMs are decreased below their optimum level, an increase in unwanted consequences of failure will occur with virtual certainty. I have witnessed this phenomenon firsthand.

Figure 9.7 also sheds some light on the immediate, as opposed to the longer-term, issue of resources. As PMs decrease and once PdM activities are fully embedded, a possible excess of resources may be temporarily evident but this excess will be short-lived since emergent CM work will begin to increase. You must be mindful not to hastily jettison any resources that will ultimately be needed in a relatively short time.

The resources needed at time (B) will ultimately be less than those needed at time (A) simply because of the trade-off of intrusive work for PdM. However, at some point an overhaul or replacement will still have to be done but not until it is truly necessary rather than at some arbitrary periodicity. Any resource savings at time (B) can be used for backlog work or to allow for an orderly attrition of those resources as a result of retirements, transfers, or the like.

9.14 Chapter Summary

- Monitoring aggregate performance criteria allows you to be proactive in quantitatively assessing your plant reliability.
- Reliability was defined by Nowlan and Heap as "the probability that an item will survive to a specified operating age under specified operating conditions without failure." This is a correct definition in regard to the reliability of a specific component or a specific system, based solely on failures.
- In addition to Nowlan and Heap's definition, I view reliability as *the cumulative and integrated rate of unwanted aggregate*

events per unit of time where the events are not limited to just equipment failures.

- The metrics used to monitor the health of a facility are analogous to those used to monitor the health of the human body.
- Many industries have paid inadequate attention to establishing an accurate and potent metric to monitor and trend plant performance and reliability.
- A compilation of subjective opinions and rhetoric should not be part of the monitoring and trending strategy.
- Reliability can be measured in quantitative terms rather than based on the beliefs or loose predictions of those responsible for compiling and managing the monitoring program.
- The metrics used in this book to create the monitoring and trending strategy include virtually all of the possible influencing events and performance data that can have an impact on plant reliability. This strategy objectively integrates all the possible dynamics that can occur, with a weighting factor incorporated for each one depending on the relative importance of the event.
- Reliability should be measured per a specific unit of time. It is more appropriately measured as a rate per 1000 unit operating hours, for example; therefore, each metric is computed for a given rate of time. The metrics should not be a static gross number of occurrences.
- The weighting factors for the metrics should be developed by a consensus of the engineering, maintenance, and operations organizations and should reflect the relative importance of each event. The number of expected events should also be developed by a consensus of these organizations.
- The performance calculations are based on a rate of time. The calculations include the expected performance rate (EPR) and the actual performance rate (APR).
- The calculations can be graphed to show a snapshot of the performance rate of the entire plant. The calculations can also be graphed to show a snapshot of the performance rate of each individual system.

- Your facility can be operating at a 99 percent capacity level and still be headed into a reliability abyss that would go undetected if a robust aggregate monitoring and trending program was absent from your strategy.
- Facilities that continuously raise the bar each year, striving for even better performance, and that embrace a very strict EPR, are among the better-functioning plants that others should try to emulate. RCM is a perfect fit for their objectives.
- A smoothly and efficiently operating plant can fall into oblivion if its management becomes complacent. Going from being a "leader of the pack" to the "bottom of the pack" does not happen overnight; it usually occurs gradually, over a period of years.
- Plant reliability was not intended to be put on autopilot. It needs constant attention and involvement.
- At some point the PMs and CMs will settle into a state of equilibrium (optimum state in Figure 9.7). Once PMs are decreased below their optimum level, an increase in unwanted consequences of failure will occur with virtual certainty.
- With any newly implemented RTF philosophy, PMs will decrease rapidly; thus worker-hours will decrease rapidly and CMs will begin a rapid ascent for approximately 12 to 18 months. Worker-hours will then begin to stabilize.
- The optimum state may shift relatively, but the same relative equilibrium prevails. The optimum state is unique for each facility, but in relative measures it is similar to others. It is like a discrete signature for each plant. The goal is not to have zero CMs; it is to seek the optimum level of equilibrium between PMs and CMs that is specifically applicable to your facility.
- At some point an overhaul or replacement will still be required but not until it is truly needed rather than at some arbitrary periodicity.

Chapter 10

RCM Implementation Made Simple—Epilogue

Congratulations! You have just completed your journey toward learning how simple it is to develop a premier classical RCM program. You have absorbed quite a bit of information, and you will no doubt be inclined to revisit specific chapters in detail for a review of the finer points of each step of the process.

10.1 RCM as a Plant Culture

Throughout this book, I have occasionally mentioned how management mind-sets and behaviors can influence the culture at a facility. By *management culture,* I mean the projected beliefs, passion, and sincerity that management exhibits and cultivates throughout the organization, in either a positive or a negative way. In my numerous speaking engagements, in meeting with industry leaders at different conferences, in visits to other facilities, and from firsthand experience, I have seen the effects of both good and bad cultures.

It is a distinct but intangible quality that distinguishes a good culture from a not-so-good culture. The roots of an organizational culture are found in the honest internal answers to such questions as these: Does your management believe in covering up anything that is negative? Is there a hidden agenda that promotes "looking good" no matter what happens? Does your management maintain only a lackadaisical belief in reliability? Is superficial "lip service" paid to reliability issues, or perhaps a "flavor of the day" mentality exists? I have seen all of these philosophies in action, and none of them work very well. No matter how hard a management team tries to cultivate an image of caring and self-assessment in regard to reliability, if the hearts and minds of that management team are not fully committed, their insincerity and shallowness will quickly be seen for what they are—smoke and mirrors. Eventually, the smoke clears and only the mirrors remain.

In contrast, I have seen cultures where management was truly committed to achieving reliability success, and it showed in their actions, their candor, and their willingness to fundamentally address weaknesses with more than mere empty words. An attitude of denial, defensiveness, slick cover-up presentations, and other gimmickry were nonexistent. A positive culture is fostered when a management team directs its efforts toward doing the right thing rather than expending the same amount of effort to circumvent what needs to be done. Those facilities that channel their effort into a genuine push for reliability, even if it is initially painful, will reap the rewards of a good culture.

Those individuals in responsible positions at your facility who have the authoritative capability to establish a truly sincere culture of reliability and preventive maintenance have a golden opportunity to advance that culture using RCM. The best-run facilities and the plants that are the shining examples of sound reliability practice are found to have in common a management culture that is enlightened, sincere in its beliefs, and ever vigilant about maintaining a reputation for being the best and safest plant it can be.

As you have learned, RCM is not just for the maintenance department. It is not just for operations. And it is not just for engineering. RCM can be the catalyst for promoting a plant culture that channels the energies of all parties toward a common goal: a safe and reliable facility. That plant culture can be obtained with the collective involvement of all stakeholders—maintenance, operations, engineering, and the craft personnel and technicians—working together with an understanding of the simple logic and commonsense approach that RCM offers. When applied correctly and simply, RCM is a universally accepted practice that has withstood the test of time and has successfully held up against an array of challenges to its straightforward logic.

As for the culture of safety, I am an absolute advocate when it comes to personnel and plant safety. As I mentioned in Chapter 1, I have had the opportunity to work in what are perhaps two of the most rigid industries in regard to safety standards: the commercial aviation and the commercial nuclear power industries. Compromising safety is never an option, and I have never lost sight of the fact that my decisions could affect the lives of grandparents, sons, daughters, husbands, wives, and other loved ones who were flying faster than 600 miles per hour at an altitude in excess of 30,000 feet in a very sophisticated and complex aircraft. Nor have I ever forgotten how mistakes in a nuclear environment potentially affect the lives of thousands of people living in an area surrounding a nuclear power plant. Adherence to safety is mandatory.

10.2 A Step-by-Step Review of the Process

As an overall review of Chapters 1 through 9, Figures 10.1*a, b,* and *c* will briefly revisit each step of the process of *RCM Implementation Made Simple*. You may wish to go back to a specific chapter for a more in-depth review.

10.2.1 Select an RCM point of contact

This is where you start. You need to select a person who has leadership and communication skills to be your RCM single point of contact (SPOC). This person should assemble at least one representative each from maintenance, operations, engineering, and selected members of the craft. The representatives can be rotated from within the various organizations and the different crafts, based on the applicable expertise needed at that point of your analysis. All team members should become familiar with *RCM Implementation Made Simple*.

Rather than prescribe some rigorous training program using facilitators and consultants, the SPOC should have the respon-

252 Chapter Ten

```
┌─────────────────────────────────┐
│ SELECT AN RCM POINT OF CONTACT  │
│ AND REPRESENTATIVES FROM MTCE,  │
│ OPS, ENGRG, AND THE CRAFT       │
└─────────────────────────────────┘
                │
                ▼
┌─────────────────────────────────┐
│ REVIEW THE REASONS FOR RCM      │
│ PROGRAM FAILURES IN CHAPTER 2   │
└─────────────────────────────────┘
                │
                ▼
┌─────────────────────────────────┐
│ BECOME INTIMATELY FAMILIAR WITH │
│ RCM IMPLEMENTATION MADE SIMPLE  │
│ ESPECIALLY THE CONCEPTS         │
│ DESCRIBED IN CHAPTER 3          │
└─────────────────────────────────┘
                │
                ▼
┌─────────────────────────────────┐
│ DEFINE YOUR ASSET RELIABILITY   │
│ CRITERIA                        │
└─────────────────────────────────┘
                │
                ▼
┌─────────────────────────────────┐
│ ESTABLISH YOUR ALPHA-NUMERIC    │
│ EQUIPMENT DATABASE TO INCLUDE   │
│ ALL PLANT I.D.'S                │
└─────────────────────────────────┘
                │
                ▼
┌─────────────────────────────────┐
│ ANALYZE EACH COMPONENT          │
│ FUNCTION IN THE "COFA"          │
│ LOGIC TREE                      │
└─────────────────────────────────┘
                │
                ▼
                        CONTINUED ON
                        FIGURE 10-1(b)
                                  ───▶
```

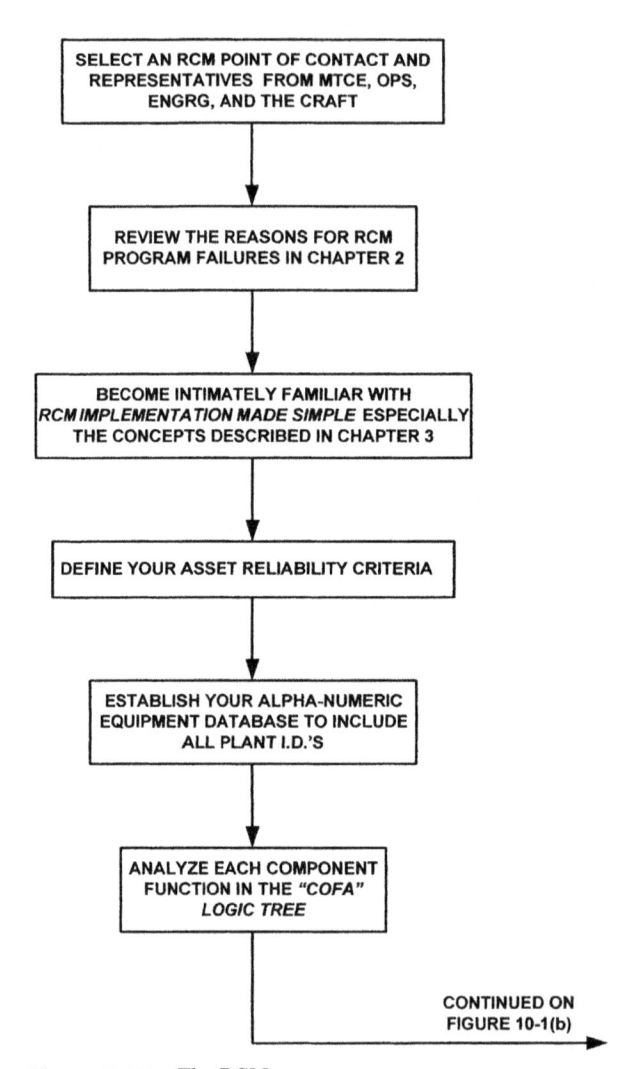

Figure 10.1(a) The RCM process.

sibility for determining how to ensure that all key players have the requisite RCM knowledge. I recommend, as a minimum, that all of the people involved in your RCM program read Chapters 3, 5, and 6. These chapters by themselves should be sufficient training for understanding RCM and being able to implement a premier reliability program.

RCM Implementation Made Simple—Epilogue 253

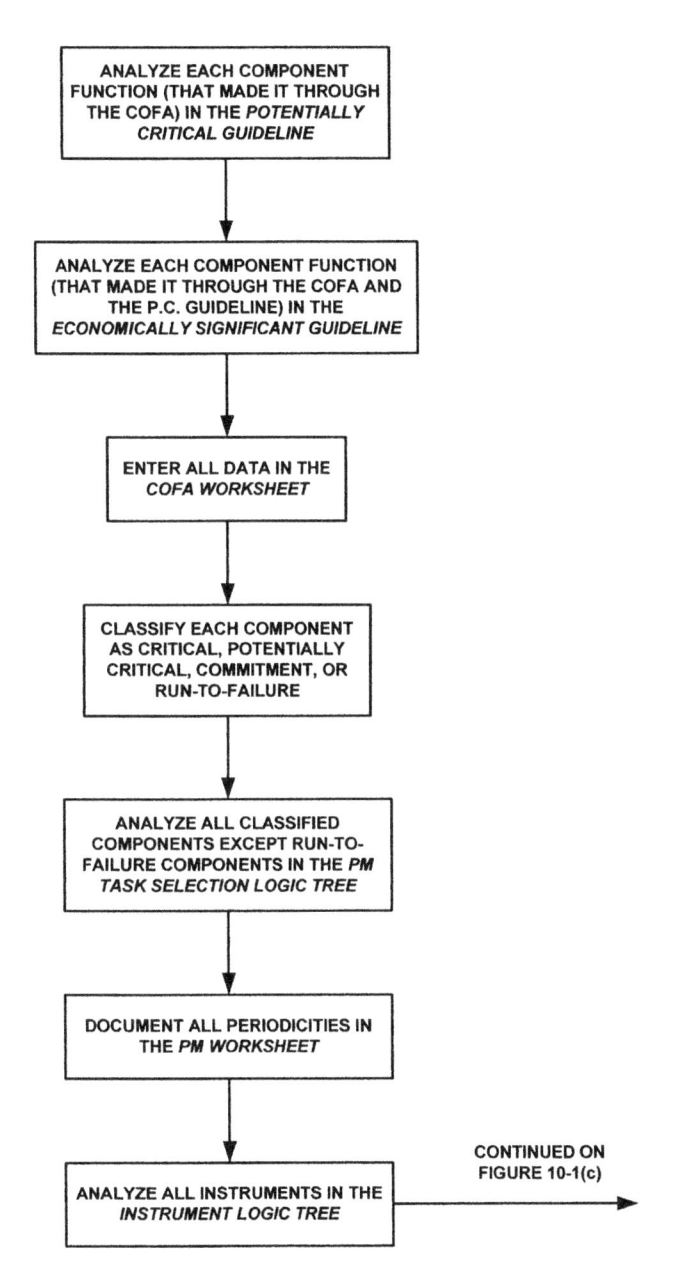

```
┌─────────────────────────────────┐
│   ANALYZE EACH COMPONENT        │
│   FUNCTION (THAT MADE IT THROUGH│
│   THE COFA) IN THE POTENTIALLY  │
│   CRITICAL GUIDELINE            │
└─────────────────────────────────┘
                │
                ▼
┌─────────────────────────────────┐
│   ANALYZE EACH COMPONENT FUNCTION│
│   (THAT MADE IT THROUGH THE COFA│
│   AND THE P.C. GUIDELINE) IN THE│
│   ECONOMICALLY SIGNIFICANT      │
│   GUIDELINE                     │
└─────────────────────────────────┘
                │
                ▼
┌─────────────────────────────────┐
│   ENTER ALL DATA IN THE         │
│   COFA WORKSHEET                │
└─────────────────────────────────┘
                │
                ▼
┌─────────────────────────────────┐
│   CLASSIFY EACH COMPONENT       │
│   AS CRITICAL, POTENTIALLY      │
│   CRITICAL, COMMITMENT, OR      │
│   RUN-TO-FAILURE                │
└─────────────────────────────────┘
                │
                ▼
┌─────────────────────────────────┐
│   ANALYZE ALL CLASSIFIED        │
│   COMPONENTS EXCEPT RUN-TO-     │
│   FAILURE COMPONENTS IN THE PM  │
│   TASK SELECTION LOGIC TREE     │
└─────────────────────────────────┘
                │
                ▼
┌─────────────────────────────────┐
│   DOCUMENT ALL PERIODICITIES IN │
│   THE PM WORKSHEET              │
└─────────────────────────────────┘
                │
                ▼                    CONTINUED ON
┌─────────────────────────────────┐  FIGURE 10-1(c)
│   ANALYZE ALL INSTRUMENTS IN THE│ ──────────────►
│   INSTRUMENT LOGIC TREE         │
└─────────────────────────────────┘
```

Figure 10.1(b) The RCM process (*continued*).

254 Chapter Ten

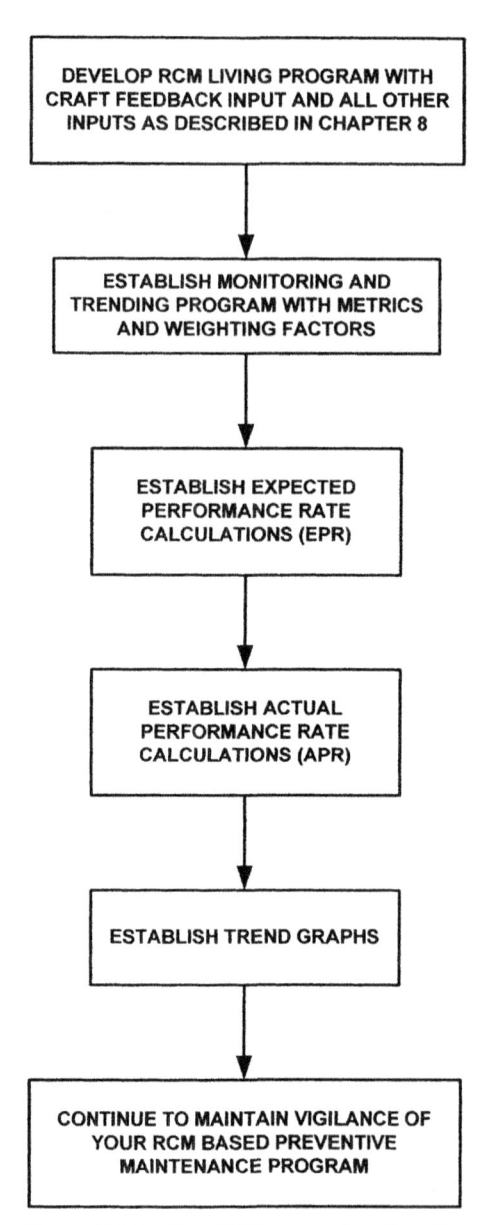

```
┌─────────────────────────────────┐
│ DEVELOP RCM LIVING PROGRAM WITH │
│ CRAFT FEEDBACK INPUT AND ALL OTHER │
│ INPUTS AS DESCRIBED IN CHAPTER 8 │
└─────────────────────────────────┘
                │
                ▼
┌─────────────────────────────────┐
│ ESTABLISH MONITORING AND        │
│ TRENDING PROGRAM WITH METRICS   │
│ AND WEIGHTING FACTORS           │
└─────────────────────────────────┘
                │
                ▼
┌─────────────────────────────────┐
│ ESTABLISH EXPECTED              │
│ PERFORMANCE RATE                │
│ CALCULATIONS (EPR)              │
└─────────────────────────────────┘
                │
                ▼
┌─────────────────────────────────┐
│ ESTABLISH ACTUAL                │
│ PERFORMANCE RATE                │
│ CALCULATIONS (APR)              │
└─────────────────────────────────┘
                │
                ▼
┌─────────────────────────────────┐
│ ESTABLISH TREND GRAPHS          │
└─────────────────────────────────┘
                │
                ▼
┌─────────────────────────────────┐
│ CONTINUE TO MAINTAIN VIGILANCE OF │
│ YOUR RCM BASED PREVENTIVE       │
│ MAINTENANCE PROGRAM             │
└─────────────────────────────────┘
```

Figure 10.1(c) The RCM process (*continued*).

10.2.2 Review the reasons for RCM program failures

This is an area that the SPOC can explain to the team. You do not want to fall into the same traps, ditches, and sinkholes that have swallowed others in their attempts to implement RCM programs. In Chapter 2, I listed the most common reasons for failing to successfully implement an RCM program. One major ingredient of such failure is the involvement of consultants. Not only are they expensive, but most of them have less knowledge of RCM than you have now that you have completed this book.

10.2.3 Understand the concepts

Understanding the concepts of *RCM Implementation Made Simple* is perhaps the single most important element in the successful implementation of your RCM program. A sure pathway to success lies in understanding the concepts of hidden failures; critical components; single-failure and multiple-failure analyses; potentially critical components, which have been the missing link of RCM; the canon law of run-to-failure; the consequence of failure analysis (COFA); redundancy; backup; and standby functions.

I cannot overemphasize the importance of understanding these concepts and applying them appropriately. RCM is fundamentally a black-and-white, commonsense decision process once you understand the ground rules and concepts. All of these concepts are explained in great detail in Chapter 3.

Remember, too, that *plant reliability and safety are related to the vulnerabilities that have* not *yet been identified because the failure consequences surrounding those vulnerabilities have not yet occurred.*

RCM Implementation Made Simple is all about finding those vulnerabilities *before* they occur and result in an unwanted consequence of failure.

10.2.4 Define your asset reliability criteria

This step in the process is where you determine your own individual plant reliability "gold standard." It is the standard you set for maintaining the reliability of your facility. The asset reliability criteria you establish should be a clear and concise definition of the unwanted events that you wish to avoid. It is what you base your entire RCM logic on. Defending your facility against these unwanted events is what your preventive maintenance program strategy is designed for. The RCM logic explores the consequences of equipment failures for their impact on any of the asset reliability criteria you selected.

Because of the importance of selecting these criteria, it is strongly recommended that the management heads of all stakeholder organizations sign a "memorandum of understanding" attesting to their concurrence, including any "qualifying conditions." Figure 5.1 lists some typical asset reliability criteria. Different industries and different types of facilities may have different criteria goals; however, the first two criteria relative to personnel and plant safety are mandatory, nonnegotiable criteria for any industry or plant. The remaining criterion covers those unwanted operational events that you and management have decided must be avoided.

10.2.5 Establish your alphanumeric equipment database

This is the database that includes all of the components in your plant. It is at the equipment level where a discrete component I.D. number is defined. These are the components that will be analyzed within your RCM COFA Logic Tree for their functions. Analyzing all equipment I.D.s is a requirement, but doing so allows you to eliminate the tedious, resource-intensive, time-consuming task of creating system and subsystem boundaries and interfaces. Piece parts such as bearings, shafts, or springs are not considered components with an equipment I.D. Piece parts become important when defining the specific causes of failure at the equipment level. Refer to Chapters 4 and 6.

10.2.6 Analyze each component function in the COFA Logic Tree

The COFA Logic Tree is where each function of the component is analyzed for its consequence of failure. The functions are the explanation for why the component is installed, and preserving these functions is the main objective of the maintenance program. The functions describe what the component must accomplish, and it is within the COFA Logic Tree that those functions are analyzed for the effects of their failures.

The COFA Logic Tree is the step of the process that identifies critical components based on their functional failure consequences. The COFA Logic Tree further determines whether a critical component failure will result in an unwanted *personnel or plant safety* consequence or an *operational* consequence. The COFA Logic Tree process will also guide you through the Potentially Critical Guideline and the Economically Significant Guideline if the component is determined not to be a critical component. Refer to Figure 5.2a.

10.2.7 Analyze each component function in the potentially critical guideline

If the first stage of the COFA Logic Tree determined that the functions of a component were not immediately critical, the functions are then analyzed within the Potentially Critical Guideline to determine whether a functional failure, in combination with additional multiple failures, an additional initiating event, or time could result in a critical plant effect. If it could, the component is classified as a potentially critical component. A commitment component is also identified within this guideline. Refer to Figure 5.2b.

10.2.8 Analyze each component function in the economically significant guideline

If the component was not determined to be critical, potentially critical, or a commitment, it is analyzed within the Economically Significant Guideline to determine whether its failure entails an economic concern. The failure of an economic component will only result in labor and/or material costs. Refer to Figure 5.2b.

10.2.9 Enter all data in the COFA worksheet

All of the data is entered in the COFA Worksheet. This includes each component I.D., the different functions of each component, the functional failures, whether the indication of failure is evident, and the consequences of failure as determined by the COFA Logic Tree, the Potentially Critical Guideline, and the Economically Significant Guideline. Refer to Chapter 5.

10.2.10 Classify each component

Each component is classified as critical, potentially critical, commitment, economic, or run-to-failure on the COFA Worksheet.

10.2.11 Analyze all classified components except run-to-failure components in the PM task selection logic tree

This is the step where the PM tasks are specified for those components that were determined to be included in the PM program—that is, critical, potentially critical, commitment, and economic components. The PM tasks will be condition-directed, time-directed, or failure-finding. Predictive maintenance tasks are a subset of the condition-directed category. Condition-directed tasks are the first choice of preventive maintenance activity rather than intrusive time-directed tasks such as overhauls, internal inspections, and replacements. Failure-finding tasks are specified for potentially critical components whose failures are hidden. Refer to Chapter 6 and Figure 6.3.

10.2.12 Document all tasks and periodicities on the PM task worksheet

All of the PM tasks and their periodicities are documented on the PM Task Worksheet. Organizations other than maintenance, such as operations and engineering, also perform PM activities.

Another important point is that establishing PM periodicities that do not have mandatory time intervals specified by a regulator is more of an art than a science. Refer to Figure 6.5.

10.2.13 Analyze instruments in the instrument logic tree

There are two instrument categories: functional instruments and instruments used for indication only. Instruments that provide a function such as a level switch or a pressure switch are analyzed in the COFA. Instruments that provide an indication only are analyzed within the Instrument Logic Tree. Refer to Chapter 7 and Figure 7.1.

10.2.14 Develop your RCM living program

Because RCM is not a static, one-shot attempt to develop a preventive maintenance program, it must remain a living and viable program, constantly being monitored for changes.

You have learned how to develop a premier RCM program based on the knowledge and awareness of the information, facts, and data that you had at the time of implementing the RCM program. This is especially true of the periodicities you prescribed for the PM tasks. However, everything is subject to change. These changes can come about as a result of new failure modes that become apparent but were nonexistent at the time of implementation. You may have experienced changes in maintenance methods that have been sufficiently beneficial to allow periodicities to be extended. On the other hand, some changes in equipment parameters may have resulted in the need for reducing periodicities to maintain reliability. Modifications to the plant, modifications to equipment, changes in operating characteristics, new PdM technologies, which are appearing everyday—all these contribute to the need for periodically reviewing and updating your maintenance program.

Your RCM living program consists of many elements. One of the most significant is the craft feedback element, which uses the expertise and involvement of the crafts personnel/technicians to provide insight into the PM tasks that have been established. Refer to Chapter 8 for a detailed review of the living program elements.

10.2.15 Establish monitoring and trending program metrics

A strategy needs to be in place that monitors and measures *aggregate* performance criteria so that you can be proactive about quantitatively assessing your plant reliability. An aggregate philosophy affords you a more in-depth measurement of your true underlying performance. It will also allow you to make quicker decisions as your program is being implemented instead of having to wait until after you have completed your program and then reactively responding to make any major course corrections if they are needed.

You need to have a measurement standard based on a *rate of time* by which to measure your performance. Crude metrics such as counting the number of times your facility incurs an unwanted trip is one metric but not the only one. The metrics in Chapter 9 are as objective as possible and easy to acquire, and they include all pertinent factors that provide you with a true picture of how well you are performing.

Each metric carries a weighting factor to differentiate its relative importance. Even though the weighting factors are arbitrarily established, as long as they remain constant, their relative importance measurement remains constant. Refer to Chapter 9 for a detailed review of a model for a monitoring and trending program.

10.2.16 Establish your expected performance rate

As part of your monitoring and trending program, you need to establish what your expectations are, and this is achieved by establishing your expected performance rate (EPR) based on the expected number of occurrences per unit of time for each individual metric. A nominal unit of time is per 1000 unit operating hours computed on a quarterly basis. Refer to Chapter 9.

10.2.17 Establish your actual performance rate

The second measurement in your monitoring and trending program is your actual performance rate (APR). This measures your performance based on the actual number of occurrences of each metric. Refer to Chapter 9.

10.2.18 Establish your trend graphs

The results from the calculations of your expected performance rate and your actual performance rate are graphed to show either a positive or a worsening trend. These graphs are extremely useful for providing senior management with a snapshot of the aggregate of all the metrics. It is a quick means to provide a very perceptive and intuitive look into the reliability window of your facility.

10.2.19 Maintain continued vigilance over your program

At some point in time, your PMs and CMs will find a state of equilibrium. This is shown in Figure 9.7 as the optimum state.

Indiscriminately tweaking a premier reliability program is not unlike adding an extra measure of salt or sugar to a proven recipe. It can totally upset the balance of that recipe, turning it from a connoisseur's delight to something inedible.

Unfortunately, a sense of smugness about being a good performer often pervades senior management meetings. This is the point at which they believe they have achieved reliability nirvana and begin to allow their facility to slip backward because they cease being vigilant about their preventive maintenance program. In my experience, this is a very common occurrence and should most definitely be avoided.

Plant reliability was not intended to be put on autopilot. It needs constant attention and involvement. The absence of unwanted plant consequences for a mature plant means that someone has been doing something right. It is usually at these peak performance times that management attempts to tweak the process and begins experimenting to further reduce more costs by eliminating what they perceive to be excess resources because the plant is operating so well. This road to reliability oblivion is more apt to appear when a new management team arrives and their first order of business is to improve the bottom line.

When you finally achieve a respectable reliability performance for your plant, the maintenance program can become a target-rich opportunity for corporate bean counters to begin slashing costs. That would be a major mistake, but it's one that, sadly, occurs with too much regularity.

Familiarize yourself with Figure 9.7 and Chapters 8 and 9. Stay continually involved and maintain constant vigilance about your RCM program!

10.3 Taking Command of Your Own Ship

I sincerely hope that *RCM Implementation Made Simple* has been an enlightening experience for you and that it will put you on a success path toward establishing a premier RCM-based preventive maintenance program. This book should have helped you to realize that implementing a classical RCM program does not require the costly assistance of consultants. It does not require any special training or facilitator intervention. It does not require the creation of system boundaries and interfaces. It does not even require an engineering degree.

Throughout this journey, I have tried to remove all of the confusion, obfuscation, and complexity from a process that should not be confusing, obfuscated, or complex. It is my goal that this book be used as an influence on setting a new universal standard for every industry that wishes to improve its plant reliability by developing an RCM program.

RCM should not be kept a secret, such that only large corporations with megabucks to spend on consultants are the beneficiaries of this invaluable process. There are many paths to improving reliability, but so far no other process has proven to be as effective as the RCM methodology for establishing the best preventive maintenance program attainable.

You now should have the tools to implement your own program and, like the captain of a ship, be in total control of the decisions and the direction you wish to travel. It has been my pleasure to lead you to the dock, assist you in getting ready to take over the helm, and enable you to sail toward a successful destination where you are in total command of your own ship!

Neil B. Bloom

Glossary

I have been encouraged by my colleagues to prepare a comprehensive RCM glossary that is more than the usual quick-phrase definitions found in most RCM-related books and references that in many instances are inadequate to explain or describe an RCM term. I frequently use examples as part of the definition in order to more clearly explain a term.

Run-to-Failure (RTF) is a typical example in which the existing simplistic definitions found in most RCM books are far too superficial to explain the true meaning of RTF. A definition of RTF such as "the component is allowed to fail with no PMs needed" is totally inadequate.

Furthermore, the definitions I use are those that are applicable to the RCM process. Obviously, Webster's dictionary offers many other definitions for the same words, but the more specific definitions that do not pertain to the RCM process are not considered.

APPLICABLE AND EFFECTIVE This term applies to each PM task. In order to be "applicable and effective," the proposed task must be appropriate for addressing the respective failure mechanism. Based on a principle of prudent judgment by knowledgeable individuals, the task should be pertinent and bear the likelihood that it will prevent the failure mechanism. The "applicable and effective" task must offer some degree of assurance that it will either prevent the failure or at the very least minimize the exposure to the failure, or minimize the exposure to a plant effect if it is a failure-finding task.

ASSET RELIABILITY CRITERIA The first step in implementing your RCM program is to define precisely, accurately, and exactly the asset reliability criteria your senior management wishes to preserve. To define these criteria, the following concerns are considered: ensuring personnel and plant safety; preventing any unplanned production delays, unplanned facility shutdowns, power reductions, production interruptions, or the loss of generation or production capacity; preventing any unwanted regulatory or environmental issues from bringing unwelcome publicity or litigation, and so on. It is strongly advised that a formal internal memorandum of understanding be signed by all applicable individuals in authority for agreement on what asset reliability criteria you choose. Your asset reliability criteria are fundamentally the identification of all the unwanted consequences of failure that can occur to your facility and that must be prevented.

BACKUP COMPONENTS   See STANDBY COMPONENTS.

BACKUP FUNCTIONS   See STANDBY FUNCTIONS.

BOUNDARIES   Ordinary classical RCM programs require the establishment of system boundaries. Boundaries comprise several hundred specific components that have to be identified when performing the analysis at the system level. Identifying functions at the system level requires that each system be specifically defined to demarcate where one system ends and another system begins. The system must be further dissected to include subsystems. An individual component can reside in only one subsystem of one system. Defining these boundaries is totally arbitrary and extremely time consuming. *RCM Implementation Made Simple* does not require the establishment of system boundaries.

CANON LAW FOR RUN-TO-FAILURE   A run-to-failure component is designated as such solely because it is understood to have no safety, operational, commitment, or economic consequence as the result of a single failure. Also, the occurrence of the failure must be evident to operations personnel.

As a result, there is *no proactive preventive maintenance strategy* to prevent failure. However, once failed, an RTF-designated component *does have a proactive corrective maintenance strategy* commensurate with all other components based on the plant conditions at that time.

CAUSES OF FAILURE   See CREDIBLE FAILURE CAUSES.

CLASSICAL RCM   This refers to the only truly acknowledged form of RCM as it was intended by its pioneers in the commercial aviation industry. Unfortunately, in its transfer from the airlines to other industries, classical RCM took on unnecessary administrative burdens and a self-imposed cumbersome process of analysis and implementation. These self-imposed restrictive impediments to successfully completing an RCM program have led to high-risk, streamlined versions of the process that are not considered classical RCM.

The streamlined versions do not meet the specificity of the SAE requirements for an RCM program that are described in SAE Document JA1011. The goal of *RCM Implementation Made Simple* is to bring classical RCM back to its simpler and more understandable roots so that it can be implemented successfully.

COFA   The COFA is a brand-new term that stands for the *consequence of failure analysis*. This format is more accurate, simpler to use, and more clearly understood than the commonly used FMEA (failure modes and effects analysis). The FMEA determines the consequence of failure, but it does so by invoking the requirement to establish functions at the *system and subsystem level*. System and subsystem functions in turn require the creation of boundaries and interfaces. The COFA establishes functions at the *component* level. Boundaries and interfaces are not required. The primary objective of RCM is to define the consequence of failure. The COFA does this directly and more accurately than the existing FMEA.

COFA LOGIC TREE   The COFA Logic Tree takes a complex logic process and simplifies it to its basic elements while enhancing its thoroughness, accuracy, and conceptual clarity. It integrates all of the logic for identifying critical, potentially critical, commitment, and economic components. See ECONOMICALLY SIGNIFICANT GUIDELINE, POTENTIALLY CRITICAL GUIDELINE.

COFA WORKSHEET   The COFA Worksheet is the format where all of the pertinent analysis is documented as decisions are made from the COFA Logic Tree, the Potentially Critical Guideline, and the Economically Significant Guideline.

COMMITMENT COMPONENTS   Commitment components are those components that have certain regulatory, environmental, OSHA, insurance, or other commitments that must be maintained, thereby requiring a preventive maintenance strategy to preclude a component from failing and causing a commitment to be missed or resulting in an infraction of the commitment. Some typical examples of commitments governing certain components are insurance commitments required for major pieces of equipment, state code commitments required for pressure vessels, EPA commitments related to environmental impacts or fluid or gaseous releases, OSHA commitments required for personnel safety, and federal regulatory agency commitments such as FAA or NRC information notices and bulletins. *Usually,* a commitment component is also associated with the component being classified as either critical or potentially critical because of its importance.

COMMON MODE FAILURES   Common mode failures are failures of a population of equipment all of which are subject to the same failure mode. There are actions that can be taken to minimize the exposure to this occurrence. It would be a prudent practice to institute a sampling program that periodically inspects one or two of the components of the population to ensure that all is well *before* any common mode problems arise, allowing very little time to react appropriately and efficiently with a planned course of action. A sampling program that performs an in-depth internal inspection of the sampled components can yield a very accurate prognosis of wear patterns, internal flaws, or other incipient failure mechanisms that may not be observable with PdM techniques alone.

COMPANION EQUIPMENT Companion equipment is those components associated with a critical or potentially critical component. Companion equipment could be an inlet or discharge check valve, a component providing an input signal, or a component that supports one of the functions of the critical or potentially critical component.

These companion equipment components should already have been independently analyzed themselves to be either critical or potentially critical. However, when viewed as companion equipment, it is more likely that they will not inadvertently be overlooked. Since all components will be analyzed regardless, it is a good check to ensure that those components associated with a critical or potentially critical component are also analyzed appropriately.

COMPONENT The terms *equipment* and *component* are synonymous in this book. This is the level at which an individual equipment I.D. number is assigned. It does not include piece parts of mechanical components such as bearings, shafts, seals, or piece parts of electrical components such as resistors, capacitors, or diodes. Piece parts are included when analyzing the causes of failure of the equipment or component.

COMPONENT CLASSIFICATIONS Components identified as being part of the preventive maintenance program will fall under one of the following classifications: *critical, potentially critical, commitment,* or *economic.* See COMMITMENT COMPONENTS, CRITICAL COMPONENTS, ECONOMIC COMPONENTS, POTENTIALLY CRITICAL COMPONENTS.

COMPONENT FAILURE The failure of a component occurs when the component no longer provides either its design or its desired function. Note that the failure can be defined by a regulatory requirement specifying that it must meet certain design operating conditions. In the absence of a specific regulatory or operating requirement, the failure of a component to provide its desired function is defined by the plant owner. The owner defines how much deviation from its like-new performance capability is acceptable. See FUNCTIONS.

CONDITION-DIRECTED TASK This is one of the three categories of preventive maintenance. Condition-directed tasks normally include tasks that measure, monitor, or analyze the condition of a component to determine whether it is operating acceptably or is about to fail. A condition-directed task does *not* constitute run-to-failure status.

Condition-directed maintenance means "don't overhaul or replace it until its condition indicates the need for overhaul or replacement." Predictive maintenance techniques are used to determine the condition of the equipment so that required overhaul or replacement can be scheduled to preclude the occurrence of a functional failure. See FAILURE-FINDING TASK, TIME-DIRECTED TASK.

CONDITION MONITORING   This is another term for *predictive maintenance*. See PREDICTIVE MAINTENANCE TASKS.

CONSEQUENCE OF FAILURE   The consequence of failure is what happens when a component fails. Will the failure affect personnel or plant safety? Will it affect the operability of the facility? Will it result in significant material costs? The consequence of failure is what the RCM process is all about. Everything you do in RCM and every part of the analysis is driven to obtain the answer to only one question: *what is the consequence of failure?* The COFA Logic Tree and the COFA Worksheet guide you through the process to determine the consequence of failure for each component function.

Once the consequence of failure is determined, it will define either a critical, a potentially critical, a commitment, an economic, or a run-to-failure component. The next strategy is to figure out how to prevent the unwanted failure consequence by means of a prescriptive preventive maintenance program that includes a vast selection of PM tasks from which to choose. See COFA, COFA LOGIC TREE.

CONVENTIONS   Conventions provide a common and consistent description of certain terms. They are used to describe the different PM task categories and types. They are also used to describe the different kinds of failure modes. Conventions are whatever you determine them to be. Normally used conventions for describing PM task categories are *condition-directed, time-directed,* and *failure-finding* (See CONDITION-DIRECTED TASK, FAILURE-FINDING TASK, TIME-DIRECTED TASK.) *Predictive maintenance* (see PREDICTIVE MAINTENANCE TASKS) is the convention used to describe nonintrusive predictive-type tasks such as vibration monitoring, thermography, or oil analysis.

To avoid confusion, the conventions for some of the commonly used failure modes are presented here:

ACTUATES INADVERTENTLY   This term describes that the component actuates when the proper signal or condition is not present.

ACTUATES PREMATURELY   This term describes that the component is out of adjustment or sticks or that some other failure is present that causes the component to indicate the proper condition when that condition is not yet present.

FAILS CLOSED This term covers three conditions: (1) component fails closed, (2) component fails to remain open, and (3) component fails to open.

FAILS OPEN This term covers three conditions: (1) component fails when it is open, (2) component fails to remain closed, and (3) component fails to close.

FAILS TO ACTUATE This term is used to indicate that the component does not operate when the proper signal is received.

FAILS TO FILTER This describes the failure of a filter or strainer to allow bypass flow or that it has holes, tears, or other damage that prevents the desired filtering effect.

FAILS TO POSITION This failure indicates that when a signal is received, the component position will not change.

RESTRICTS FLOW This describes the condition when a filter or a strainer becomes either partially or completely clogged and does not allow the required flow to pass through.

CORNERSTONES OF RCM The three cornerstones of RCM are (1) know when a single-failure analysis is acceptable and when it is not acceptable, (2) know how to identify hidden failures, and (3) know when a multiple-failure analysis is required.

CORRECTIVE MAINTENANCE Corrective maintenance is a strategy to fix components once they have failed. A total proactive maintenance plan includes corrective maintenance as well as preventive maintenance as an integral part of its strategy. These two entities are performed integrally to prevent a failure consequence at the plant level.

CRAFT FEEDBACK This is one of the most significant elements of the living program. It provides direct feedback from the craft personnel performing the PM tasks that were established. The craft feedback element allows for the expert opinion and best judgment of the craft personnel or technician to determine the appropriateness of the PM task. This includes the work scope of the task, the task periodicity, and whether there is a recommendation for some other modification or adjustment to the task.

CREDIBLE FAILURE CAUSES Credible failure causes are those causes of equipment failure that are most likely to occur and should not include a list of all theoretical or postulated causes that in actuality may never occur. Pertinent experience in combination with prudent judgment on the part of knowledgeable individuals is a necessary criterion in ascertaining credible failure causes.

The causes of failure include failures of the piece parts, for example, bearing failures, seal failures, motor winding failures, limit switch failures, gear failures, and shaft failures.

CRITICAL COMPONENTS Critical components are those components for which the occurrence of the failure is *evident* and the failure *immediately* results in an unwanted plant consequence. Critical component failures are at the top of the RCM component classification hierarchy, having the top priority for preventing a failure because their failure consequence is immediate. See POTENTIALLY CRITICAL COMPONENTS.

DEMAND MODE OF OPERATION Components that do not normally operate and are primarily used in standby safety systems and backup functions are analyzed in their demand mode of operation as though they were called upon to function. The component's function is normally inactive or "not in use" when the system is in service and becomes active only when an event or signal occurs that triggers the demand for the component's functional activation.

DISASTERS Disasters can be caused by acts of nature, human error, or equipment failures. Acts-of-nature disasters such as hurricanes, earthquakes, tsunamis, tornados, and landslides do not lend themselves to being tamed by human intervention, so for the most part they are unavoidable. There may be warning systems such as tsunami warning buoys in the Pacific Rim or construction standards that may help to prevent structures from buckling during an earthquake, but the event itself is unavoidable. On the other hand, human error, such as pilot error, judgment error, or operator error, offers latitude in circumventing the potential for inducing a disaster.

Disasters that happen in factories, plants, or other facilities usually have their origin in the failure of equipment. These types of failures probably offer the greatest latitude of all for circumventing their potential to induce a disaster. Nothing is ever 100 percent reliable, whether it is an aircraft, a space shuttle, or a nuclear power plant. However, disasters caused by equipment failure have the capability to be harnessed to the degree that allows for the closest proximity to that 100 percent reliability threshold. That cannot be said for natural disasters or human-induced ones.

DOMINANT FAILURE MODE For each functional failure there is a failure mode. The dominant failure modes are those modes of failure that are most likely to occur. For example, a failure mode of a valve or switch would be that the valve "fails closed" or the switch "fails to actuate." Note that the dominant failure modes are not the same as the failure *causes*. The valve could fail closed because of hinge pin wear, for example. The hinge pin wear is the cause of the dominant failure mode.

ECONOMICALLY SIGNIFICANT GUIDELINE Components that are not classified as critical, potentially critical, or commitment could be economically significant. The Economically Significant Guideline governs the decision process for the component being classified as economically significant.

ECONOMIC COMPONENTS Economically significant components are those components whose failure results in an economic consequence only. The consequence of failure results only in labor and/or material costs. There is no safety or operational concern with the failure of an economic component. If the failure of a high-cost component also resulted in a safety or operational concern, it would default to the highest classification, which would be critical, potentially critical, or commitment.

EQUIPMENT See COMPONENT.

EVIDENT FAILURE In RCM terminology, an evident failure refers to the ability of the operating crew, during their routine duties, to observe that a failure has occurred. Either the failure of the component itself is evident or the consequence of the failure is evident to the operating crew during their routine duties. See EVIDENT INDICATION.

EVIDENT INDICATION This is an indication of the failure of the specific component, or an indication of related failure consequences involving the component, that becomes evident. The indication of failure must be evident to operators in the control room, in the auxiliary control stations, or during formal, proceduralized operator rounds.

FACILITY *Facility* and *plant* are synonymous in this book. They are the entities for which you should maintain a reliability program. A facility or a plant could be any entity—a cruise ship, the space shuttle, an off-shore oil drilling platform, a manufacturing or production facility, an aircraft, an electrical power station, the electrical grid network, a water treatment plant, a hospital, a refinery, an assembly line, a paper mill, or any other complex entity requiring a reliability program.

FAILURE See COMPONENT FAILURE.

FAILURE CAUSE The causes of failure involve the piece parts of the equipment. For example, the bearings could be a cause for a pump or motor failure, the butterfly disc could be the cause for a valve failure, the internal diaphragm could be the cause for an air-operated valve failure, and so on. The preventive maintenance strategy is designed to prevent functional failures by specifying PM tasks that target the causes of those functional failures in an effort to prevent them from occurring. See CREDIBLE FAILURE CAUSES.

FAILURE EFFECT See CONSEQUENCE OF FAILURE.

FAILURE-FINDING TASK This is one of the three categories of preventive maintenance. Failure-finding tasks are specified when a condition-directed or time-directed task is not applicable. A failure-finding task will not prevent a failure at the component level. Rather, a failure-finding task is part of the preventive maintenance strategy to determine that a hidden failure has occurred. In doing so, the failure-finding task can avert an unwanted consequence at the plant level before an additional failure or initiating event occurs in combination with the hidden failure. See CONDITION-DIRECTED TASK, TIME-DIRECTED TASK.

FAILURE MODE See DOMINANT FAILURE MODE.

FAILURE MODES AND EFFECTS ANALYSIS (FMEA) This is the accustomed standard logic matrix for classical RCM, which requires that functions be defined at the system and subsystem level. This results in the need to establish system boundaries and interfaces. In *RCM Implementation Made Simple,* the FMEA is replaced by the consequence of failure analysis (COFA). See COFA, COFA LOGIC TREE.

FAILURE MODES AND EFFECTS AND CRITICALITY ANALYSIS (FMECA) The FMECA is similar to the FMEA, except it also includes the identification of the causes of failure at the same time as functions and failure effects are being analyzed. In *RCM Implementation Made Simple,* the causes of failure are analyzed separately within the PM Task Logic Tree and the PM Worksheet. This is to provide greater efficiency in the process of selecting the different types of PM tasks because PM tasks can be grouped together by component type. See PM TASK SELECTION LOGIC TREE, PM TASK WORKSHEET.

FREQUENCY Frequency is the specification of the base of time used to determine the periodicity. The frequency is normally described in terms of the following time base: hour, day, week, month, and year. The frequency is used in conjunction with the interval to define a specific periodicity. See INTERVAL, PERIODICITY.

FUNCTIONAL FAILURES Typically, the functional failures are the exact opposites of the functions. For example, if the function is "to provide a flow path for cooling heat exchanger X while pump Y is operating," the functional failure will be "fails to provide a flow path for cooling heat exchanger X while pump Y is operating." See also FUNCTIONS.

FUNCTIONS A function describes what the component must accomplish. The function is the explanation for why the component is installed, and preserving these functions is the main objective of the maintenance program.

In the absence of a regulatory requirement that specifically defines the operational parameters that the function must meet, the only functional definition you need to specify is a *performance standard* at a level determined by you, the owner of the facility, and the owner of your RCM program. For example, in the absence of any other specific requirements, the function of a pump could be written as: "provides the necessary flow needed to maintain the tank inventory at its nominal level."

HIDDEN FAILURES Hidden failures do not result in an immediate plant effect since, by definition, for a plant-affecting situation to occur, the failure or the result of failure must be evident to operating personnel. If the effects of a failure are not observable, the failure will have no immediate impact. When a component is required to perform its function and its consequence of failure is not evident—that is, the immediate overall operation of the system remains unaffected in either the normal or the demand mode of operation—the failure is a hidden failure. The purpose of identifying hidden failures is to prevent the exposure to plant effects that could result from multiple failures, time, or other initiating events involving the respective component in its hidden failure state.

HIDDEN FUNCTIONS A component has a hidden function when either of the following conditions exist:

1. The function is normally active or in use when the system is in service, but there is no indication to operating personnel when the function is lost or ceases to perform.
2. The function is normally inactive or not in use when the system is in service and there is no indication to operating personnel that the function is not available when needed.

See also HIDDEN FAILURES, STANDBY COMPONENTS, AND STANDBY FUNCTIONS.

INDICATION OF FAILURE See EVIDENT FAILURE, EVIDENT INDICATION.

INITIATING EVENTS An additional event that can occur such as a loss of on-site power, flooding after a storm, a local fire, or any other event that could, in combination with an existing failure, result in an unwanted plant consequence. See POTENTIALLY CRITICAL COMPONENTS.

INSTRUMENT LOGIC TREE This logic tree guides you through the process of identifying the preventive maintenance strategy for instruments.

INSTRUMENTS Instruments can be either functional instruments, those that provide a specific function such as a level switch that controls another valve, or they can be indication-only instruments that provide only an indication reading from a gage or monitor, for example, and do not in themselves provide a direct function.

INTERFACES Interfaces are auxiliaries outside the system or subsystem that are required for operation of the system or subsystem being analyzed. Interfaces provided by systems outside the system are designated as *out-system in-interfaces*. Interfaces provided by the analyzed system or subsystem that are needed to operate outside systems are designated as *in-system out-interfaces*. Interfaces provided by another subsystem within the same system are designated as *in-system in-interfaces*.

INTERVAL The interval is an integer used with the frequency to define a specific periodicity. For example, if a PM task is accomplished every two months, the frequency would be monthly and the *interval* would be 2, which would read "M2." The periodicity is therefore M2. If the task is accomplished every six months the interval would be 6 and the periodicity would be M6. See FREQUENCY, PERIODICITY.

LIVING RCM PROGRAM Because RCM is not a static, one-shot attempt to develop a preventive maintenance program, it must remain a living and viable program, constantly being monitored for changes.

Your RCM program was based on the knowledge and awareness of the information, facts, and data that you had at the time of implementing the program. This is especially true of the periodicities prescribed for the PM tasks. However, everything is subject to change. These changes can come about as a result of new failure modes that become apparent but were nonexistent at the time of implementation. Changes in maintenance methods may be sufficiently beneficial to allow periodicities to be extended. On the other hand, some changes in equipment parameters may have resulted in the need for reducing periodicities to maintain reliability. Modifications to the plant, modifications to equipment, changes in operating characteristics, and new PdM technologies, which are appearing everyday, all contribute to the need for continuously reviewing and updating your maintenance program and maintaining it as a living program.

MISSING LINK The missing link of RCM has been the absence of a definitive category for component failures that are hidden and not immediately evident, which are referred to as *potentially critical components* (i.e., having the potential to become critical). See POTENTIALLY CRITICAL COMPONENTS.

MONITORING AND TRENDING This is a strategy that monitors and measures aggregate performance criteria to allow for quantitatively assessing plant reliability. An aggregate philosophy affords the capability to have an instantaneous snapshot of the reliability of the plant and the effectiveness of the preventive maintenance program. The monitoring and trending strategy shows the trended relationship between the expected performance rate (EPR) and the actual performance rate (APR).

MULTIPLE FAILURES A multiple-failure analysis is required when the occurrence of a single failure is hidden. In the event of a hidden failure, an unwanted consequence may not occur until additional multiple failures occur. Multiple-failure analysis is one of the three cornerstones of RCM. Even in the case of a run-to-failure component where the occurrence of the failure is evident, the failure must be corrected in a timely manner before additional multiple failures occur. See CORNERSTONES OF RCM, POTENTIALLY CRITICAL COMPONENTS, CANON LAW FOR RUN-TO-FAILURE.

NONCREDIBLE FAILURE A noncredible failure has theoretical or postulated causes that in actuality may never occur or are very unlikely to occur. See CREDIBLE FAILURE CAUSES.

NORMAL MODE OF OPERATION The normal operating mode for a component or system is when the function of that component or system is normally active. The normal mode of operation represents the ability to discern, during routine plant operation, that a component is functioning properly or that it has failed. See DEMAND MODE OF OPERATION.

OPERABILITY CONCERNS These are consequences of failure that pertain to the operability of the facility, and are distinct from safety concerns and economic concerns. See ASSET RELIABILITY CRITERIA, ECONOMIC COMPONENTS.

OPERATOR ROUNDS Operator rounds are an integral part of any preventive maintenance program. These are the routinely scheduled, formalized inspections made by the plant operating staff to ensure that plant equipment is operating properly. They offer the ability to identify equipment that may be making uncharacteristic noises, or to identify oil, water, steam, or other fluid leaks. Operator rounds provide the capability for routinely monitoring expected pressures, temperatures, electrical readings, and so on, of plant equipment.

PERIODICITY The periodicity includes both the frequency *and* the interval. For example, an A2 periodicity indicates an annual frequency A *plus* the interval of 2, meaning the task is performed every two years. Using the frequency alone can be misleading. For example, extending the frequency from weekly to monthly means you perform the task *less often,* thereby increasing the periodicity. Conversely, reducing the frequency from monthly to weekly means you perform the task *more often,* thereby decreasing the periodicity.

PHASES OF RCM

Phase 1: Consists of *identifying* equipment that is important to plant safety, generation (or production), and asset protection.
Phase 2: Consists of *specifying* the requisite PM tasks for the equipment identified in phase 1. These tasks must be both applicable and effective.
Phase 3: Consists of properly *executing* the tasks specified in phase 2.

PIECE PARTS These are the individual parts that are responsible for the causes of equipment failures. See CREDIBLE FAILURE CAUSES.

PLANT See FACILITY.

PM TASK SELECTION LOGIC TREE This contains the logic for selecting the PM category of either a time-directed, a condition-directed, or a failure-finding task. It also provides the logic for discerning when a design change is necessary.

PM TASK WORKSHEET This is the worksheet on which decisions are documented relative to specifying the various PM tasks. It is applicable only to those components that were identified as critical, potentially critical, commitment, or economic.

POTENTIAL FAILURES This term should not be confused with *potentially critical*. Potential failures are incipient equipment failures that are usually detected by predictive maintenance techniques such as vibration analysis, oil sampling, and thermography. A potential failure refers to the detection of the impending or imminent failure of a component. It is a component that is about to fail—for example, a component that is vibrating, is running at a higher than normal temperature, has a fluid leak, or is making unusual noises.

POTENTIALLY CRITICAL COMPONENTS This component classification has been the missing link of RCM. A potentially critical component is one whose immediate failure is *not evident* and is *not immediately* critical but has the potential to become critical either with a duration of time in and of itself, with an additional failure, or with an additional initiating event, at which time the consequence of the failure may unfortunately become quite evident (and critical). The potential to become critical can occur not just with an additional component failure, but also with an additional initiating event, or even with an additional routine plant evolution such as turning on a seldom-used system, switching on a newly modified circuit, or shutting down a pump under certain conditions.

Potentially critical components can be thought of as sleeper cells lying in wait to wreak havoc upon your plant or facility. They are failed components that are not evident, so no one knows they have failed. They are components that no longer perform their function, but you don't and won't know about their loss of function and potential consequence of failure until an additional failure, initiating event, or evolution occurs, causing the sleeper cell to manifest itself. This is a vulnerability that must be avoided!

The majority (approx 98 percent) of potentially critical components are so designated as a result of the effects of multiple failures or initiating events, as opposed to the effects of time duration (approximately 2 percent). The reason for including the aspect of time duration is for the analysis to be absolutely complete, thorough, and accurate so that nothing escapes the analysis or is inadvertently omitted.

Potentially critical components as a result of multiple failures or initiating events: When two (or more) components (valves, pumps, motors, etc.) operate to supply a function that each can fulfill individually, and there is no indication of failure for each component individually, then a failure of one of the components is hidden (there is no indication the component has failed), and the failure does not result in an immediate plant effect. However, if a second component should fail, or if some other initiating event or plant evolution takes place that would otherwise rely on the failed component, then a plant-effecting consequence would occur. Hence, the component is considered *potentially* critical. For example, if there are two pumps normally operating at the same time, a failure of the pump discharge check valve in the open position will be hidden. Only when the associated pump fails will the unwanted reverse flow path through the failed open check valve become evident.

Potentially critical components as a result of time: A typical example of being potentially critical due to time is if one panel of a multipanel circulating-water traveling screen, which filters seaweed and other ocean debris, were to fail or become damaged, there would be no indication that the panel had failed, nor would there be an immediate effect. However, over a duration of time, the failure of one of the screen's panels, *in and of itself,* can eventually cause clogging of the heat exchangers by failing to filter debris, which will ultimately result in a plant effect. Note that the traveling screen never fails completely or immediately and that you do not need to have a second additional failure for this consequence to occur. Another example of classifying a component as being potentially critical because of time duration is a large tank that has a small leak that has not been detected. This would not be evident immediately, nor would it require a second multiple failure to manifest itself. However, over a given duration of time, in and of itself, a plant consequence may occur, since the inventory of the tank represents an important function. These are vulnerabilities that you will not know about until they have already resulted in a plant consequence—which is not the time to find out about them. When that happens, it is too late to take preventive action.

POTENTIALLY CRITICAL GUIDELINE This guideline contains the logic for determining when a component is classified as being potentially critical. It represents an extremely important facet of an RCM program. See POTENTIALLY CRITICAL COMPONENTS, RCM FILTER.

PREDICTIVE MAINTENANCE TASKS Predictive maintenance (PdM) is a subset of condition-directed maintenance and includes using mostly nonintrusive technologies to monitor equipment for precursors to failure. Predictive maintenance techniques are used to determine the condition of the equipment so that incipient failures can be detected and required overhaul or replacement can be scheduled to preclude the occurrence of a functional failure.

Commonly used PdM techniques include vibration analysis, thermography, oil sampling and analysis, acoustic monitoring, radiography inspection, magnetic particle inspection, eddy current testing, ultrasonic testing, liquid penetrant, motor current signature analysis, boroscope inspections, motor-operated valve diagnostic testing, and air-operated valve diagnostic testing.

PREVENTIVE MAINTENANCE Preventive maintenance is the strategy designed to prevent an unwanted consequence of failure. This strategy could be directed at preventing failures at the component level or it could be designed for preventing failures directly at the plant level. Preventing failures at the component level is obviously a well-understood concept.

Preventing failures directly at the plant level, however, includes specifying failure-finding tasks and including corrective maintenance as part of the overall preventive maintenance strategy. Failure-finding tasks are necessary to preclude the occurrence of an unwanted plant consequence as the result of a hidden failure in combination with additional multiple failures or other initiating events. Corrective maintenance, too, must be addressed for an overall preventive maintenance strategy to preclude the occurrence of an unwanted plant consequence as the result of failure of a run-to-failure component in combination with additional multiple failures or other initiating events. Including RTF components as part of the overall preventive maintenance strategy is virtually always overlooked.

Preventive maintenance includes specifying condition-directed, time-directed, or failure-finding tasks to prevent failure consequences. As part of these three main categories, predictive maintenance tasks, operator rounds, and a whole host of other types of tasks are prescribed. See CONDITION-DIRECTED TASK, CONSEQUENCE OF FAILURE, FAILURE-FINDING TASK, PREDICTIVE MAINTENANCE TASKS, TIME-DIRECTED TASK.

QUALIFYING CONDITIONS (FOR ASSET RELIABILITY CRITERIA)
When selecting the asset reliability criteria, which are the consequences that must be avoided to ensure the protection and productive capability of your facility, qualifying conditions can sometimes be placed on the criteria as a threshold point for allowing a certain amount of latitude before specifying that an unwanted consequence has actually occurred. For example, in the event of an unplanned power reduction, the threshold may be 30 minutes, meaning that as long as the power reduction was for *less* than 30 minutes, the unplanned power reduction threshold would not be met. See ASSET RELIABILITY CRITERIA.

RCM FILTER This is a depiction of the filter displaying the uncomplicated decision logic process of *RCM Implementation Made Simple*. It illustrates how every component must pass through the different stages of this filter to discern their classification. The first stage of the filter determines whether the component is critical, the second stage of the filter determines whether the component is potentially critical, the third stage identifies commitment components, and the fourth stage defines economically significant components. Any component that makes it through all four filter stages is classified as run-to-failure. See COMMITMENT COMPONENTS, CRITICAL COMPONENTS, ECONOMIC COMPONENTS, POTENTIALLY CRITICAL COMPONENTS, RUN-TO-FAILURE (RTF). See also Figure 5.3.

REDUNDANCY Redundancy is a widely misunderstood term. Redundancy means more than just having two components instead of one component supplying a function, in the belief that preventive maintenance is not that important in such a situation. Although *redundancy*, according to its most basic definition, does mean that there is more than one component available to perform a particular function, redundancy alone does not provide sufficient indication that one of the two (or more) redundant components has failed.

Standby and backup components are considered to offer redundancy, but they do not operate simultaneously. Truly redundant components *usually* operate simultaneously. Redundancy in general, however, does not imply that there is an indication of failure for each redundant component. See STANDBY COMPONENTS, STANDBY FUNCTIONS.

RELIABILITY   Reliability is the probability that an item will survive to a specified operating age, under specified operating conditions, without failure. More comprehensively, reliability can be defined as the cumulative and integrated rate of unwanted aggregate events per unit of time, where the events are not limited to just equipment failures. Reliability includes a whole host of unwanted events and occurrences that can be measured as a rate of time. Reliability represents a broader spectrum of events than just failures; therefore, reliability measurements, not just limited to failures, can offer much more intuitive insight for determining the effectiveness of your preventive maintenance program and how well your facility is being run.

RELIABILITY-CENTERED MAINTENANCE (RCM)   RCM comprises a set of tasks generated on the basis of a systematic evaluation that are used to develop or optimize a maintenance program. RCM incorporates decision logic to ascertain the safety and operational consequences of failure and identifies the mechanisms responsible for those failures.

RUN-TO-FAILURE (RTF)   RTF is the misunderstood orphan of reliability. Most people, engineers included, provide the automatic response that if a component fails and nothing happens, it is a run-to-failure component. This is completely wrong. Another prevalent but totally incorrect assumption is that having redundant components or redundant systems automatically means the component or system is run-to-failure. The common definition of RTF found in most RCM books will simply state that "the component is allowed to fail with no PMs needed." That definition is far too shallow to accurately define RTF and prevent mismanagement of it. A very precise and prescriptive means of identifying when a component can be classified as run-to-failure is needed. I have called this the *Canon Law for Run-to-Failure*. See CANON LAW FOR RUN-TO-FAILURE.

RUN-TO-FAILURE ANOMALIES   If the component is low enough in the hierarchy of relative importance, although the component failure might be evident, there still may not be any plant consequence even if there were additional failures in conjunction with the original failure. When the logic delineates a multiple-failure scenario that still does not have any unwanted consequence of failure, always ask, "Why is the component even installed in the plant?" Another possible anomaly would exist for those components that are installed in your facility strictly for convenience or that have insignificant value. It is not uncommon to find that even though such components have failures that are not evident, in accordance with the process logic, they may still end up being classified as RTF.

The inherent benefit of the canon law for run-to-failure is that it makes these anomaly components stand out and be noticed, so that no component of importance escapes the RCM logic. It provides a path for exception, but only after that exception has been carefully analyzed. Any anomaly components can then be evaluated for whether they should continue to be maintained or whether they should be considered for removal from the plant entirely—but that decision is yours.

SAE STANDARD FOR RCM The Society of Automotive Engineers (SAE) developed a standard that entitles a process to be called an RCM process. The main reason for this was that the term *RCM* was being applied to a multitude of PM program enhancements that had no technical logic and were not systematically developed. They were a conglomeration of PM betterment efforts, or PM review efforts, or PM program updates, that were improperly described under the pretext of being called RCM. The SAE wanted to ensure the isolation of these other, rather arbitrary, efforts from the more defined and thorough approach of specifically applying RCM logic. In consequence, they issued a fundamental standard referred to as SAE Document JA1011 that had to be met in order to call the maintenance program process an RCM process.

SAMPLING This is a strategy that performs a comprehensive internal inspection of specific components that have been in operation for many years without ever having been overhauled or replaced, and may have undetected incipient failure mechanisms taking place. There may be strong suspicions that certain major components need to be overhauled or replaced, because pump seals, impellers, stator windings, and bearings don't last forever, even if you have been religiously performing routine predictive maintenance tasks such as monitoring for excessive vibration and sampling and changing the oil. It is still not an easy feat to convince senior management to buy into a widespread overhaul plan without evidence of equipment degradation, and that is where a sampling strategy can be of immense reliability value.

Another reason for performing a sampling inspection is to confirm the validity of your predictive maintenance technologies. It is not uncommon for major failure mechanisms to go undetected on some equipment, even with a host of PdM tasks that are regularly accomplished.

SINGLE-FAILURE ANALYSIS RCM is nominally a single-failure analysis except when the failure is hidden. It then becomes a multiple-failure analysis. See HIDDEN FAILURES, HIDDEN FUNCTIONS, POTENTIALLY CRITICAL COMPONENTS.

SLEEPER CELL   This expression is used to describe how a hidden failure remains undetected, similar to being in a sleeper cell mode, waiting until an additional failure or initiating event occurs, which in combination with the hidden failure causes an unwanted plant consequence.

STANDBY COMPONENTS   These components are described as either standby, backup, or redundant components. Standby and backup components are usually not operated simultaneously. They operate in a standby or backup mode of operation in the event of failure or upon the demand of another component. They may or may not be similar components. A standby or backup pump, for example, may activate on the demand of a pressure or flow switch. They can function automatically on an input signal, or they can function by manual initiation. Unlike standby or backup components, truly redundant components or systems are usually like components or systems that operate simultaneously.

STANDBY FUNCTIONS   These functions are usually associated with normal standby and standby safety systems. The system must be analyzed in its demand or functional operating mode to identify critical and potentially critical components, since the system is not normally operating. Standby systems must have surveillance or failure-finding tasks specified on a periodic basis to ensure the operability of the system or component so it can function as intended to prevent a consequence of failure.

STREAMLINED RCM   This includes various forms of preventive maintenance that are truncated versions of classical RCM. For the most part they do not meet the definition of an RCM program as defined by the Society of Automotive Engineers in accordance with SAE Document JA1011. See SAE STANDARD FOR RCM.

TIME-DIRECTED TASK   This is one of the three categories of preventive maintenance. Time-directed tasks are usually intrusive maintenance activities such as overhauls, disassemblies, replacements, and internal inspections, which are scheduled according to a given periodicity. See CONDITION-DIRECTED TASK, FAILURE-FINDING TASK.

TIMELY MANNER   This is the commonly used term for describing when corrective maintenance on a run-to-failure component must be performed. A run-to-failure component must be corrected in a timely manner depending on the plant conditions at that time. The timely manner may be measured in hours, days, or weeks depending on the judgment and the decision process of operations and engineering personnel and on the existing plant conditions. A timely manner cannot under any circumstance be limitless. See CANON LAW FOR RUN-TO-FAILURE, CORRECTIVE MAINTENANCE, RUN-TO-FAILURE (RTF), RUN-TO-FAILURE ANOMALIES.

Bibliography

Author's Note: I have limited the bibliography to only those sources that I consider pertinent to understanding RCM. There are, of course, many other good books and publications on RCM that I am familiar with and could list, but for the most part they do not offer any additional insight into RCM. The reference documents I have listed cover the fundamental issues of RCM.

Air Transport Association of America. Airline/Manufacturer Maintenance Program Planning Document, Maintenance Steering Group (MSG)—2, 1970.

Air Transport Association of America. Airline Program Development Document, Maintenance Steering Group (MSG)—3 Task Force Document, 1980.

Bloom, Neil. "Reliability Centered Maintenance—A Case Study." Presented at the Argonne National Laboratory in cooperation with the United States Nuclear Regulatory Commission, Argonne, IL, 1992.

Bloom, Neil. "The Role of RCM in the Overall Maintenance Program." Presented at the International Atomic Energy Agency (IAEA), Vienna, Austria, 1991.

Bloom, Neil. "Understanding Hidden Failures in RCM Analyses." *Maintenance Technology,* 2003.

Bloom, Neil. "Understanding Hidden Failures in RCM Analyses," Presented at the Process and Power Plant Reliability Conference, Houston, TX, 2003.

Moubray, John. "Is Streamlined RCM Worth the Risk?" *Maintenance Technology,* 2001.

Moubray, John. *Reliability Centered Maintenance RCM II,* 2nd ed., Industrial Press, 1997.

Netherton, Dana. "RCM Tasks," *Maintenance Technology,* 2004.

Netherton, Dana. "Standard to Define RCM," *Maintenance Technology,* 2003.

Nowlan, F. Stanley, and Howard F. Heap. *Reliability Centered Maintenance.* U.S. Department of Commerce, National Technical Information Services, Report Number AD/A066-579, December 1978.

Index

Age, relationship to failure, 162–163, 178
Aggregate metrics, 10, 217–223, 239–240, 260–261
  as part of performance graphs, 237
  weighting factors for, 230
"Applicable and effective," 30, 67, 80–82, 117, 158–160
  instrument-related, 182
  versus *non*applicable and effective, 169
  in phase 2 of RCM, 131
  and PM Worksheet, 100
  when it cannot be defined, 112
Asset reliability criteria, 30, 50, 80–82, 100, 209, 255–256
  defining, 107, 109–119, 129–132, 139, 144
  as part of aggregate metrics, 225
  as part of COFA, 157
  as part of PM Task Worksheet, 157
  typical examples of, 111
  (*See also* Qualifying conditions)

Backup components and functions (*see* Standby components and functions)
Bathtub curve, 161–162, 244
Benchmarking, 239
Boundaries, 3, 256, 262
  associated with RCM made "difficult," 147–149
  confusion concerning, 21–22
  eliminating requirement for, 75–78, 98–99
  unnecessary, 28, 34, 92

Cannon law for run-to-failure, 4, 25, 27, 52–58, 68–70, 86, 255
  definition of, 54–55
Causes of failure, 79, 87, 94, 102, 116, 153–154, 158–162, 167–168, 256
  (*See also* Credible failure cause; Piece parts)

Classifications (*see* Component classifications)
COFA, 87, 107–108, 146–147, 208–210, 214, 255–259
  compared to SAE Standard, 134
  defined, 4, 28
  as first stage of RCM filter, 119–120
  versus FMEA, 81–82, 95
  for instruments, 182, 185, 190
  integrated with PM task selection logic, 153, 157, 169
  for plant design changes, 208
  reason for, 82, 90–100
  understanding of, 112–115, 119–140
COFA Logic Tree, 82, 124, 137–139, 146, 256–258
  diagram of, 113
  as first stage of RCM filter, 119–121
  understanding of, 112–115, 119–140
  (*See also* COFA)
COFA worksheet, 82, 97, 100, 105, 108, 121–132, 137, 146, 157, 258
  for instruments, 185
  for plant design changes, 208
  understanding of, 112–115
  (*See also* COFA; COFA Logic Tree)
Commitment components, 26, 50–55, 74–75, 82, 86, 146, 160, 178, 257–258
  in living RCM program, 204–210
  in PM strategy, 50–51, 74, 100, 112
  related to implementation process, 112–119
  related to instruments, 186
  as third stage of RCM filter, 119
Common mode failures, 43, 171, 187
Companion equipment, 133
Component(s), 6–7, 17–18, 21–22
  alphanumeric database of, 91–92
  analyzing, 77–78, 93
  assigned at specific I.D. number, 93
  classifications of, 73–74
  default, 67
  grouping of, 34

287

Component(s) (*Cont.*):
  inappropriate classifications of, 25–26
  as sole level of importance, 22
  staying at level of, 34, 42, 45
  subtle classification differences of, 40
  synonymous with equipment, 78–79
  (*See also* Commitment components; Critical components; Economic components; Potentially critical components; Run-to-failure components)
Component classifications, 56, 72–74, 110, 113, 129, 131
  examples of, 72–73
  hierarchy of, 73–74
Concepts of RCM, 3–4, 8, 13–14, 18, 55–56, 68–69, 85–86, 255
  for averting disaster, 61–65
  explained for first time, 27–34
  if misunderstood, 21
  pertaining to hidden failures and redundancy, 24–28
  as preparation for RCM implementation, 105
Condition-directed task, 154–156, 159–162, 172, 175–176, 205, 212, 258
  (*See also* Bathtub curve; Failure-finding task; Time-directed task)
Confusion surrounding RCM, 17, 46, 82, 262
  regarding convention, 24, 94, 154
  regarding system boundaries and interfaces, 21
  regarding system functions, 70
  regarding traditional renditions of RCM, 114
Consequence of failure, 4–6, 27–28, 85, 87, 115, 255–256
  anomalies and, 68
  and asset reliability criteria, 100, 129
  in bigger picture, 59–61
  depending on how function is written, 67
  economics of, 27
  final destination of, 82
  independent of component pedigree, 81
  for instruments, 182, 190
  potential result of, 37–39, 48
  quest for, 21, 79
  as ultimate objective of PM program, 57

Consequence of failure analysis (COFA) (*see* COFA)
Consultants, 3, 11, 13, 15–16, 95, 251, 255, 262
Convention terminology, 24, 91, 105, 269
  for failure modes, 94
  for instruments, 181
  for PM tasks, 154
Cornerstones of RCM, 3, 32
Corrective maintenance (CM), 8, 24–25, 39, 53–61, 81, 114, 157, 186
  as integral part of run-to-failure, 27, 45, 53–61, 68–69, 86
  as part of living RCM program, 195–196, 200, 203–204, 210, 213
  as part of monitoring and trending, 229, 244
Craft feedback, 10, 196–202, 210, 213, 259
Credible failure cause, 160, 167–168
  versus *non*credible failure cause, 116
Critical components, 17, 26, 34–35, 39–40, 52, 69, 85, 255–257
  beginning logic process by defining, 113–119
  deeper look at, 65
  as first stage of RCM filter, 119
  as part of component hierarchy, 73–74
  as part of defensive PM program strategy, 75
  and sampling strategy, 169–170

Demand mode of operation, 33, 36, 45–46, 67, 126, 155
  (*See also* Normal mode of operation)
Design change, 43, 64, 86, 112, 134, 141, 208, 224, 245
  and craft feedback element, 198
  for critical and potentially critical components, 119
  depending on applicable and effective task, 160
  for instruments, 184–191
  recommendation criteria for, 166–168, 204
Disasters, 2, 141
Divergent expectations, 23
Dominant failure mode, 98–99, 167–168
  defined, 116, 146
  described for each functional failure, 124–130, 138

Economic components, 51, 75, 82–83, 86, 102, 160, 178, 258
  in economic evaluation worksheet, 113
  as fourth stage of RCM filter, 119
  included in equipment database, 210
Economically significant guideline, 112–120, 146, 153, 204, 257–258
  illustrated, 114
  instrument-related, 182, 190
  as part of RCM filter, 119
Equipment (*see* Component)
Evident failures, 10, 29, 33–55, 63–70, 82, 85–86, 258
  as first question to RCM COFA decision logic, 115–118
  as part of COFA worksheet, 124–132
Evident indication (*see* Evident failures)

Facility, 12, 129
  synonymous with plant, 272
  (*See also* Types of facilities)
Failure:
  age-related (*see* Age, relationship to failure)
  causes of (*see* Credible failure cause)
  effect of (*see* COFA; Consequence of failure)
  evident (*see* Evident failures)
  hidden (*see* Hidden failures)
Failure-finding task, 43, 46, 154–156, 160, 167, 178, 258
  (*See also* Condition-directed task; Time-directed task)
Failure modes (*see* Dominant failure mode; Hidden failure mode)
Failure modes and effects analysis (FMEA):
  versus COFA, 81–82, 87
  versus COFA spreadsheet, 95–100
  as part of RCM made "difficult," 147–148
Frequency and interval (*see* Periodicity)
Functions, 12, 45, 87, 91, 93, 121, 138, 145, 256–258
  described, 121, 123–125
  grouping not allowed for, 34, 202
  not required at system level, 22, 28
  related to companion equipment, 133
  specified at component level, 34, 76–82, 98–99

Functions (*Cont.*):
  (*See also* Confusion surrounding RCM, regarding system functions; Standby functions)
Functional failures, 77, 80, 91, 99, 110, 123–124, 150, 258
  (*See also* Functions)

Health of plant, 217–231, 247
Hidden failure mode, 66
  (*See also* Hidden failures)
Hidden failures, 3–4, 24–25, 66, 255
  applicability to failure-finding tasks, 155, 178
  deeper look at, 65
  as first question to RCM COFA decision logic, 115–118
  not part of RTF philosophy, 52–55, 64–65
  related to missing link, 46–50, 85
  understanding more about, 27–42
Hidden functions, 274
  (*See also* Hidden failures)
Hidden systems, 25
  (*See also* Evident failures; Hidden failures; Hidden functions; Standby functions; Testing hidden systems)

Immediate effect, 25, 39, 46–49, 57
  (*See also* Critical components)
Initiating events, 47–50, 68, 75, 85, 155, 160
  and Potentially Critical Guideline, 114, 118, 139, 257
Instrument logic tree, 182–183, 259
  (*See also* Instruments)
Instruments, 14, 26, 87, 181–191, 259
Interfaces (*see* Boundaries)
In-house control, 7, 16–19, 95, 109

Living RCM program, 10, 13–14, 79, 165, 193–215, 259–260
  analysis element, 209
  corrective maintenance element, 203
  craft feedback element (*see* Craft feedback)
  engineering evaluations, 208
  equipment database, 210
  industry failure data, 207
  monitoring and trending, 208
  new commitments, 208
  other inputs, 205

290   Index

Living RCM program (*Cont.*):
  plant design changes, 208
  PM audit, 210
  regulatory bulletins, 207
  root-cause evaluations, 206
  vendor bulletins, 206
Living RCM program model, illustrated, 195

Misunderstanding:
  of hidden failures, 24
  of redundancy, 70
  of run-to-failure, 25
Missing link, 18, 46, 64–65, 85, 255
  (*See also* Potentially critical components*)
Monitoring and trending, 10, 195–196, 217–245, 260–261
  of equipment versus facility, 209
Multiple failures, 4, 25, 32–69, 85, 255

Normal mode of operation, 33, 45–47, 66–67, 71–73, 85–87, 126, 178
  (*See also* Demand mode of operation)
Normally active (*see* Normal mode of operation)
Normally inactive (*see* Demand mode of operation)

Operability testing (*see* Testing hidden systems)
Operator rounds, 70, 116–117, 124, 186
Operational criteria (*see* Asset reliability criteria)
Optimum state, 244–245, 248, 261

Performance, 13–14, 84, 195, 209, 214, 260
  calculations, 230–235
  comparison with human body, 220–222
  monitoring strategy, 221
  plant metrics of, 217, 219, 222–230
  weighting factors of, 230–231
  (*See also* Aggregate metrics; Monitoring and trending; Performance graphs; Performance rates)
Performance graph(s), 235–240, 261
Performance rate(s), 219, 223, 234–236, 241, 247, 260–261
  actual, 234–235, 247, 261
  expected, 234–236, 241, 247, 260–261

Periodicity, 81, 83, 153, 156, 163–165, 210, 212–213, 246, 248
  and craft feedback, 196, 198–200, 203
  defined, 9–10
  determining optimum for, 164–165
  for PdM tasks, 172
  for instruments, 182–191
  related to grace period, 230
Phases of RCM, 8, 29–31, 85, 131, 153
Piece parts, 79, 87, 93, 102, 162, 256
  (*See also* Causes of failure)
Plant (*see* Facility)
Potential failure, of given component, 48
Potentially critical components, 4, 8, 18, 27, 29, 37, 42–58, 62–69, 73–75, 82–83, 100, 110–120, 133, 137–141, 255–258
  deeper look at, 65
  and failure-finding tasks, 160, 178
  and multiple failures, 48–49
  as second stage of RCM filter, 119
  and time, 48–49
  (*See also* Missing link; Potentially Critical Guideline)
Potentially Critical Guideline, 112–120, 139–140
  illustrated, 114
  as part of RCM filter, 119
  (*See also* Potentially critical components)
Predictive maintenance (PdM), 30, 83, 155–159, 170
  types of, 172–175
  (*See also* Condition-directed task)
Preventive maintenance (PM), 1–13, 46–47, 67–68, 107, 112, 250, 256
  and aggregate metrics, 223–229
  compared to defensive strategy of football team, 75
  and complacency, 241–246
  for instruments, 181–182
  integrated with corrective maintenance, 57–59
  and living RCM program, 194, 203–214
  maintaining vigilance of, 261–262
  as means of translating plant design objectives, 131
  monitoring of, 237–241
  and operator rounds, 70
  as part of canon law for run-to-failure, 53–57

PM task selection logic tree for, 159, 211, 258
and PM Task Worksheet, 100, 158
three phases of, 30–32
understanding PM task terminology for, 154–157

Qualifying conditions, 109–111, 146, 225, 256
(*See also* Asset reliability criteria)

RCM (*see* Reliability centered maintenance)
RCM concepts (*see* Concepts of RCM)
RCM decision logic tree (*see* COFA Logic Tree)
RCM filter, 119–120
RCM living program (*see* Living RCM program)
RCM pitfalls, 13–15, 18, 26
Redundancy, 3, 8, 17, 34–37, 43, 50, 62–64, 70–73, 255
misunderstanding of, 24–25
(*See also* Standby functions)
Reliability, 3–5, 12–13, 27, 33, 65, 193, 250–262
complacency about, 241–242
maintaining performance of, 242–248
measuring, 217–239
optimum time to establish program for, 165
Reliability centered maintenance (RCM), 1–14
classical, 5–6, 28, 76, 79, 87, 175
definition of, 8
"made difficult," 147–151
not defined as PM reduction program, 16, 23
as plant culture, 249–251
streamlined, 5–7, 141–143, 147–148
as white elephant, 16

Run-to-failure, 1, 4, 8, 25–26, 37, 52–58, 258
anomalies, 68–69
canon law for, 54–55
in component classifications, 73–74
versus economic component, 102
in equipment database, 210
as misunderstood orphan of reliability, 53, 56, 61
as part of bigger picture, 59
in RCM filter, 119–120
(*See also* Canon law for run-to-failure)

SAE Standard, 14, 29, 80–81, 99, 134–135, 147, 210
Safety criteria (*see* Asset reliability criteria)
Sampling strategy, 169–171
Single failure, 3–4, 33–37, 50–55, 62–64, 75, 85, 117, 255
Single point of contact (SPOC), 90, 108, 211, 251
Sleeper cell, 7, 37, 43–48, 64, 141
Standby components and functions, 3–4, 25, 34–40, 66–67, 70–73, 86–87, 255
analyzing, 45

Testing hidden systems, 41–42, 45–47, 62, 64–65, 85
Time-directed task, 155, 159–162, 258
(*See also* Bathtub curve; Condition-directed task; Failure-finding task)
Timely manner, 38–39, 45, 53–58, 68
(*See also* Canon law for run-to-failure; Run-to-failure)
Types of facilities, 12, 272

Unnecessary burdens, 20, 28–29, 77

Work control organization, 30–31, 204

ABOUT THE AUTHOR

Neil Bloom received his bachelor of science degree in mechanical engineering from the University of Miami, where he also minored in economics. He has worked in close association for more than 30 years with the two leading-edge federal agencies most responsible for reliability and safety, namely, the Federal Aviation Administration (FAA) and the Nuclear Regulatory Commission (NRC). His airline and nuclear work in both engineering and maintenance has afforded him unique background and knowledge as a practitioner of reliability centered maintenance (RCM). He understands what can and cannot be done with classical RCM. He knows what works and what doesn't. He recognizes the pitfalls and knows how to circumvent them.

Mr. Bloom has created totally new classical RCM concepts and philosophies that go beyond the work of the pioneers of RCM, namely, Stanley Nowlan and Howard Heap, in order to make the entire process less daunting, more straightforward, and simpler. He has been responsible for developing and managing what is perhaps, even today, one of the most comprehensive classical RCM programs ever developed. The program has analyzed more than 125,000 components at one of the country's largest nuclear generating facilities.

Mr. Bloom has been a guest speaker on RCM at some of the most prestigious national and international conferences, including the Electric Power Research Institute (EPRI), the American Society of Mechanical Engineers (ASME), the American Nuclear Society (ANS), the Argonne National Laboratory (ANL), which is operated by the University of Chicago for the Department of Energy (DOE), the Edison Electric Institute (EEI), and the International Atomic Energy Agency (IAEA) in Vienna, Austria.

Mr. Bloom and his wife Bernadette reside in Monarch Beach, California. He can be reached via e-mail at neilbloom@rcmauthor.com.

CPSIA information can be obtained
at www.ICGtesting.com
Printed in the USA
BVOW06*0019310317
479801BV00005B/53/P

9 780071 460699